Educating Israel

Educating Israel

Educational Entrepreneurship in Israel's Multicultural Society

Yehuda Bar Shalom

palgrave
macmillan

EDUCATING ISRAEL

Softcover reprint of the hardcover 1st edition 2006 978-1-4039-7274-3

First published in 2006 by
PALGRAVE MACMILLAN™
175 Fifth Avenue, New York, N.Y. 10010 and
Houndmills, Basingstoke, Hampshire, England RG21 6XS
Companies and representatives throughout the world.

PALGRAVE MACMILLAN is the global academic imprint of the Palgrave Macmillan division of St. Martin's Press, LLC and of Palgrave Macmillan Ltd. Macmillan® is a registered trademark in the United States, United Kingdom and other countries. Palgrave is a registered trademark in the European Union and other countries.

ISBN 978-1-349-53395-4 ISBN 978-1-4039-8361-9 (eBook)
DOI 10.1057/9781403983619

Library of Congress Cataloging-in-Publication Data

Bar Shalom, Yehuda.
[Ide'ah shel tikun. English]
Educating Israel : educational entrepreneurship in Israel's multicultural society / Yehuda Bar Shalom.
p. cm.
Includes bibliographical references and (p.) index.
Translated from Hebrew

1. Multicultural education—Israel. 2. Pluralism (Social sciences)—Study and teaching—Israel. 3. Multiculturalism—Israel. I. Title.

LC1099.5.175B37413 2006
370.117095694—dc22 2005058651

A catalogue record for this book is available from the British Library.

Design by Newgen Imaging Systems (P) Ltd., Chennai, India.

First edition: August 2006

This book is dedicated to my spouse, Amira

This book was written with kind support from

The David Yellin College of Education, Jerusalem
The Mofet Institute, Tel Aviv
Temple Israel, Tulsa, Oklahoma
The Jewish Federation of Tulsa, Oklahoma
Alan B. Slifka Foundation, New York
Lucuis N. Littauer Foundation, New York

CONTENTS

ACKNOWLEDGMENTS

I am indebted to many people who, over the years, have helped me to formulate the ideas presented in this book. I would like to thank Professor Charles Harrington and Professor Hope Jensen Leichter of Teachers College, Professor Joseph Lukinsky of the Jewish Theological Seminary, Dr. (Reverend) Earl Kooperkamp of Union Theological Seminary, Rabbi Andrew Hahn, from New York and Avraham Infeld of Melitz, Israel. Thanks to Dr. Itai Zimran, Dr. Nita Shochet, Dr. Revital Heiman, Dr. Anna Rousso, Dr. Ora Gravelsky, Dr. Ilana Sigad, Professor Aharon Shai, Dr. Bilha Piamenta, Talma Margalit, Dr. Hansaa Diab, Reuven Azulai, Sigal Muskal-Atzmon, Dr. Ruth Borstein, Dr. Dan Woolf, Dr. Yehudit Woolf, Yiftah Goldman, Dr. Einat Guberman, Dr. Tamar Asher, Dr. Amalia Shach, Yaron Lehavi, Vered Shavit, Dr. Raaya Yoeli, Dr. Revital Sela-Schiovitz, Dr. Reuven Garber, Miki Garber, Aryeh Ben David and Eyal Bloch, all of David Yellin Teachers College. I have been influenced by many other lecturers and students at David Yellin and appreciate their insightful comments on the research findings. Thanks to Dr. Yitzhak Mendelson, moderator of the Jewish-Arab dialogue group at David Yellin, which helped me clarify my thoughts in this field. Thanks to Dr. Yair Caspi, who helped me understand my role in both research and education. Thanks to Professor David Ohana of Ben Gurion University for his inspiring comments. Thanks to Dr. Zvi Bekerman and Professor Philip Wexler of the Hebrew University of Jerusalem, to Yonatan Glaser of the association Hiburim, and to Loui Dobin of the Greene Family Camp in Texas. I thank Panim for Jewish Renewal in Israel, as well as a Jerusalem businessman who wishes to remain anonymous, for their donations. Thanks to Rabbi Levi Weiman-Kelman of Kol Haneshama Congregation, and to Luli Harris for her support and assistance. Special thanks to Rabbi Charles Sherman and Perry Simons of Tulsa, Oklahoma and to Professor

Martin Belsky and Professor Paul Finkelman, both from The University of Tulsa. Thanks to Stacey Schusterman from Tulsa for her support, and to many other anonymous Tulsan donors. Thanks to Giora Rosen and to my editor, Timna Horowitz, for their wonderful help and support throughout the process of publishing the Hebrew version of this book.

During my years of work as a teacher, lecturer and researcher, I have found that the martial arts and meditation help me develop the "mental muscles" needed to cope with various challenges. These disciplines helped me cope with the challenge of writing this book. Accordingly, thanks are due to my teacher in the Ohshima Shotokan method, Eli Cohen, who is presently serving as the Israeli ambassador to Japan. Thanks to my teacher and master, Aryeh Levy, to my karate students at Givat Shaul and Reut, who have stayed with me over the years, and to the Goenka's Vipassana Association, which taught me a wonderful technique.

Thanks to my wife, Amira, and to my children Nitzan, Ran and Yehonatan. Publishing a book is a demanding process, and I thank them for their tolerance.

Thanks to Shaul Vardi (Jerusalem) who translated this book into English, helped with the English editing, and prepared the index.

Above all, I extend heartfelt thanks to the communities of educators and repairers presented in this book, and to all those working in a kindred spirit to repair Israeli society in education systems and in other spheres. I hope their voices will be heard more loudly, leading to the emergence of a better society that will adopt a humane and just social contract. This is nothing less than the basic right of each and every one of us.

PREFACE

The State of Israel was established in 1948. To a large extent, Israeli culture is still in the process of crystallization, providing a fascinating living laboratory that offers insights into the adaptive capabilities of humans in a society that is involved in developing its identity and searching for social goals.

Israeli society has not yet produced an ethos of solidarity based on a clear and common myth. The emerging behavioral patterns range from pluralism and multiculturalism, at best, to division and sectarianism, without the possibility for a unifying narrative, at worst.[1] The need to reach decisions on political, social, and cultural issues has obliged Israeli society to become aware of the modalities by which such decisions are made. The developing relations between the different communities that comprise the Israeli mosaic will clearly influence the manner and quality of these decisions.

Israeli society is characterized by a level of multiculturalism that leads to the emergence of rifts. The most prominent problems within this society are as follows:

- The conflict between Arabs and Jews
- The conflict between religious and secular Jews
- The tension between religion and democracy
- The issue of migrant laborers and their integration in Israeli society
- Ethnic stratification within the Jewish population

Within the Israeli education system, there are several formal frameworks that have assumed the task of addressing some of these conflicts. The educational ideology of the education system forms the infrastructure and justification for the existence of the educational framework, and for the manner in which it addresses the problem. Several schools around

Israel operate according to a unique ideological approach to these issues:

- The elementary school at Neve Shalom nurtures coexistence between Arabs and Jews;
- Bialik School in Tel Aviv develops an appropriate education system for the children of migrant laborers;
- Keshet School, whose guiding ideology is dialogue between secular and religious Jews;
- Kedma School, which aims to provide cultural nurturing for children from the Mizrachi community;
- The Democratic School in Hadera, which focuses on inculcating a perception of democracy.

This book examines these schools, all of which aim in different ways to mend or *Repair* existing reality.[2] In this context, this term means that these schools strive to Repair the particular rift within which they function (Arab-Jewish, secular-religious, Mizrachi, etc.); in most cases, this goal emerges against the background of a personal narrative of distress on the part of the founders. In this book, I shall describe and analyze the manner in which this ideology is manifested in each school; as we shall see, both similarities and differences will emerge among the different schools included in the study.

Case Studies

The use of case studies and descriptions is now commonplace in educational training (LaBoskey, 1992). A case study teaches educators how to act in a similar situation or when the content is similar (Merriam, 1991). Its force comes not from quantitative representation, which cannot usually be achieved in a qualitative study, but rather from the ability of those in the field to implement some of the conclusions that it offers in cases with a similar context.

To date, little attention has been paid in the professional literature to the schools discussed in this project. When documentation has been undertaken, it has usually discussed a single specific school, thus lacking a comparative dimension. In this book, I shall present each school separately, before examining their common characteristics as agents seeking to achieve the Repair of Israeli society, with its rifts and challenges.

The elementary school in Neve Shalom strives to develop a curriculum enabling a positive encounter between Jewish and Arab children.

The school addresses the difficulties inherent in an encounter between two populations involved in a protracted political, social, cultural and religious conflict. It offers a way to contain the dispute while enabling the children to engage in a real encounter with the culture of the other. Many issues at the school become the focus for argument and power struggles, such as language (the school teaches in Arabic and Hebrew), emblems, festivals and ceremonies, and, of course, the content of studies. The encounter between the two populations helps change attitudes and encourages the participants to examine themselves (Feuverger, 1995). The community of Neve Shalom is also home to the Peace School (as distinct from the elementary school discussed in this book), which organizes encounters between Jewish and Arab students from throughout Israel. The goal of these activities is to reduce prejudice and change attitudes by fostering an atmosphere of cooperation and intimacy between students from both peoples (Feuverger, 1997). There is no direct connection between the Peace School and the elementary school discussed here, though there is a clear affinity between these two institutions, which are situated in the same community and address the same subject.[3]

Bialik School, situated close to the new central bus station in Tel Aviv, attempts to meet the human needs of its students, most of whom live in very difficult conditions. Due to demographic changes, this school has been transformed from an Israeli institution to one catering to students from diverse cultures, including Arabs and migrant laborers. The school population includes the children of migrant laborers, children from the families of Palestinian collaborators who have been obliged to move to Israel, children from the families of the "South Lebanese Army," and a minority of Israeli Jewish children from native families and from immigrant families of low socioeconomic status who live in the vicinity of the old central bus station.

The problems facing the children of migrant laborers throughout the Western world, including those at Bialik School, include concern for their parents' legal status, discrimination in the host country, exposure to violence and traumatic experiences in areas of combat from which they migrated, and the poverty and neglect of urban life (Willshire, 1989; Olsen & Chen, 1987; Valdivieso, 1990). The chapter on Bialik School presents the school ethos and the adaptation of the educational perception to meet the special needs of the students. The unique approach of the school includes flexible response mechanisms on the part of the teachers, the development of multicultural sensitivity, changes and adaptations to the curriculum, and the structural changes in studies. The school functions not only as an academic institution for its students, but also as an educating institution and support community (Larivee, 2000; Pang et al., 1999), replacing and compensating for aspects that are missing in the children's homes and environments. The unique process undergone by the teachers at the school has made many of them leaders and advocates for the migrants.

Keshet School was established in order to meet the need for an encounter between secular and religious Jews. The school criticizes the decision made in the 1950s to establish separate education streams for religious and secular Jews and has developed a curriculum in which the students address issues presented in the texts from sharply differing perspectives. The school creates a space in which the children can examine their Jewish and Israeli identity. An earlier study showed that many of the parents reported experiencing changes in their way of life due to the contact they had with the school, as well as a high level of satisfaction. The school has managed to foster both a positive image of the other—secular and religious—and to blur, to an extent, the separate identities (Weil and Roer-Strier, 2000). In the school, the complexity of the identities and the blurring of the line dividing the two are reminiscent of the claim made by Selchov and other authors in Zucker (1999) that the usual labels—"secular," "Israeli," and "Jewish"—are inadequate and belie the complexity of these identities. The school plays an important role in a period when the rift between religious and secular Jews is growing (as evidenced in part by the growth of Shinui[4]), and it reacts to an unnatural situation wherein Jews from different streams meet each other for the first time (if at all) after they are called to the army.[5]

Kedma School in Jerusalem was established on the basis of an ideology that advocates the social reform of disadvantaged sections of the population. For many years, the Ministry of Education attempted to promote these populations, which are primarily of Mizrachi origin, through special curricula, but these were often of little benefit, possibly due to the structural bias and discrimination in Israeli society (Swirski, 1990, 1995). Kedma School adopted a different educational approach from that of other schools. It focuses its efforts on a specific population, in an effort to change the personal and cultural perception of its students. In this respect, it follows the approach of Taylor, who argues that there is a link between positive individual self-perception, influenced by the individual's ability to fully express their identity and culture, and their need for recognition by the environment (Taylor, 1992). The school enables the students to clearly identify the components of their identity in a critical manner and approaches Mizrachi identity not as a monolithic or structured unit, but as an important narrative that has been silenced for many years and should not be revived or even given attention. In this approach, the school combines its efforts with the efforts of movements such as the Mizrachi Democratic Rainbow that advocate the establishment of a multicultural and pluralistic society in Israel providing rights, protection, and a voice for weak and underprivileged minority groups (see Shohat, 1995; Shenhav, 1998; Criff, 1998).

The Democratic School in Hadera offers a radical educational alternative to regular Israeli schools. Of all the schools discussed in this book, the distinguishing feature of this school is that the problem it addresses is of a universal rather than uniquely Israeli character. Nevertheless, the school adds an important voice to the discourse on education and on democracy and civics in Israel—a discourse that is negligible or nonexistent within Israeli society. The school echoes the criticism of Kohn (2000) and Rogers (1973), who sharply criticize industrial and frontal educational methods, advocating holistic approaches to education that place the child at the center and emphasize the affective aspects of education; it is also consonant with the educational approach of Neil (2003) and Greenberg (1998), who established free and democratic schools as alternatives to traditional education. The school presents itself as a community that changes constantly with each new person who enters or leaves . Each voice in the community influences and changes the school. The school manages to inculcate democratic approaches to different issues among the students. For example, a survey of attitudes at the Democratic School found that its students present more democratic positions regarding the Israeli Arabs than do students in regular schools (Goldenberg, 1998).

These schools are the subject of considerable interest and discussion. They seem to radiate a spirit that has enriched educational activity in Israel. Countless delegations of educators, students and policymakers visit these schools in order to become acquainted with these innovative models and learn from their experience.

Having visited the schools many times and examined the existing literature, I have come to the recognition that these institutions constitute a genre of schools devoted to the ideal of Repair.[6] The founding ideology seeks to reduce and repair the rifts in Israeli society, expanding from this base to develop an educational practice grounded in a pluralistic and universalistic educational perspective that emphasizes the need to accept the other. Nurturing the affective side of education is particularly important in these schools. Teachers and students in these schools give each other love and attention on a much higher level than in ordinary schools. Nodding (1992) and Larrivee (2000) describe this phenomenon as a caring community, since it is based on the principles of respect, authenticity, consideration, and emotional morality. The teachers are perceived by the students as caring, attentive to their needs, eager to help, involved and interested in their success (Adler, 2002).

Although the various schools discussed in this book ostensibly depart from different starting points, considerable similarities can be found between them. The founding ideology of each school is particularistic insofar as it concentrates on repairing a specific problem, but this approach expands, creating an ideology of universal Repair.

In this book, I shall attempt to illustrate the manner in which educational theory is put into practice in schools that adhere to the "Ideology of Repair."

Methodology

I chose to employ the technique of naturalistic research, which is appropriate for examining educational processes. In each school observations were held; various documents, such as the school newspaper and the constitution, were examined in order to sharpen the understanding of the school's character and examine its messages; and a comparison was undertaken between the different schools. The criteria for this comparison were many: consonance between the declared ideology and actual educational practice, various aspects of the school culture, modalities for addressing discipline, the level of involvement of the students in educational projects, the level of involvement of the parents, and the relationship between the school and the community. During conversations with several of the principals of these schools, they complained that they do not have time to document their work properly—something that could help expand the circle of people aware of their approach, as well as facilitate a critical and reflexive examination of their work.

The research included interviews with principals, teachers, students, and parents. The observations took place during subject lessons and during activities relating to social and educational aspects. The collection of data through diverse means (interviews, observations, and documents) helped ensure a process of "triangulation."[7]

Teacher Training

There is growing awareness in the Western world of the need for educational methods to adapt to the demographic changes that are occurring in the prosperous countries. Israeli society is an immigrant society that creates an intercultural encounter between Jews and the Arab minority, as well as a growing number of migrant laborers. Although Jewish immigration to Israel (*Aliyah*) takes place under the influence of the Zionist ideology of "settling" the Land of Israel, the manner in which new immigrants integrate into Israeli society is similar to the absorption patterns of immigrants in other prosperous countries. The global phenomenon of migration has led researchers and educators to include the objective of nurturing a positive approach to multiculturalism in teacher training programs (Sleeter & Grant, 1988; Banks & Banks, 1989; Bennett, 1990).

A careful examination of the academic and educational strategies adopted by schools operating according to the Ideology of Repair will certainly contribute to an enhanced understanding of a practical theory that has developed from the grassroots in the context of repairing or addressing the cultural gaps in Israeli society. A presentation of these often-complex educational models may help Israeli teachers in the twenty-first century as they function in a multicultural society that is still only just emerging, and it may also be useful in preparing students for their entry into the field of education. This research presents models and theories that may help teacher training institutions to enrich their training programs with insights drawn from the field of educational practice.

The Narrative Base

In qualitative research, the researcher is the most important tool for understanding the research. I agree with those who argue that personal narrative can help the reader to appreciate the author's starting point, and hence understand trends, biases, and preferences in research and in writing. Accordingly, I would argue that the reader should be afforded a glimpse of such narratives as an additional basis for understanding the data. Readers interested in the findings of the study are invited to skip this section.

I was born in Argentina in 1960 to Jewish parents. My father was a communist, though one who was disillusioned with the party following the Stalin era, and was persecuted by the authorities for most of his life. My mother had a split identity, since her own father was Christian. It has sometimes been suggested in the family that, had we stayed in Argentina, we would have appeared on the lists of the Desaparecidos—the victims of the reign of terror imposed by the army generals during the 1970s. My brother and sister—who are both significantly older than me—fell in love with the Zionist idea in the late 1960s. They both came to Israel as adolescents and settled on kibbutzim. I was sent to a private Jewish school called Tarbut. I did not like the school, and I suspect that the feeling was mutual. I recall many hours in the classroom spent escaping into an imaginary world of my own thoughts. Meetings between my parents and the school counselor, dull encounters with a psychologist, and even a clinical examination of my brain waves were all to no avail. My parents, evidently despairing, decided to send me to Israel on my own at the age of eleven. At the end of the fifth grade, I was accepted as an external student on Kibbutz Machanayim through the Youth Aliyah organization. There were no children of my age on the kibbutz where my brother and sister lived, so I could not be sent there.

For me, life on kibbutz was an experience of Repair, and an opportunity to experience the socialism that my father had sometimes referred to. The free, informal approach to learning meant that, for the first time in my life, I liked school. As the years passed, however, I came to realize that I was different from my classmates. My parents were not with me, and this factor weighed against me in the eighth grade (I stayed there until the eighth grade), when a new class teacher who treated the students according to their parents' status arrived. There was no place for me in this group (the heightened sense of criticism that is characteristic of adolescence no doubt contributed to this awareness). The class teacher used to sit on my desk, her backside hiding my friends from me! I used to put thumb tacks, nails and other sharp objects on the desk to deter her, but without even looking at me she would brush these aside. For me, the way she treated me came to symbolize my status in the class and on the kibbutz.

Toward the end of the year, she called me in and made her position quite clear: nothing much could be expected of me in academic terms, but if I put my energy into working in the dairy sheds, things would be different. She wondered aloud whether I was really sure that I wanted to live on kibbutz.

At this time, my parents had just arrived in Israel and bought a pharmacy in Haifa. The conversation with the class teacher encouraged me to move back with them, although they were somewhat nervous about this.

To my amazement, in the ninth grade I became a model student at a municipal school in Haifa. The lessons were no easier than on kibbutz, but I felt that the class teacher liked me. With the help of positive reinforcement, I proved that I could learn and do well. With hindsight, my experiences at this early stage of life gave me a first-hand lesson in the potentially harmful effects of discrimination. In Argentina, we were always afraid of what the non-Jews might do—particularly those who had a Nazi past. On kibbutz, I felt rejected because I did not have the "right" family background (perhaps this explains why I later married an Israeli-born kibbutznik—a member of what I then perceived as the Israeli "aristocracy").

After completing school and military service, I was accepted to join the cinema studies track at Tel Aviv University, hopeful that I might one day become a famous and influential cinematographer. I studied in the prestigious directing program and prepared my final project together with Isaac Florentin, who now lives in Los Angeles and makes action movies. After completing my studies, I went through a period of crisis. I wanted to make a difference in the world, but realized that I could not achieve this through the field of cinema and the media. As I grappled with my doubts, my teacher Avraham Heffner spoke to us about "the influence of art." I recall his saying that it is kind of sad that directors mortgage their homes and wreck their family life in order to direct a movie that, at best, will only preach to the converted. He wondered how much better a place the world might be if all these artists invested as much in their children as they did in their artistic endeavors.

I was 25 at the time, and Heffner's views influenced me profoundly. I interpreted to his comments as referring to children in general and decided to enter the field of education. I applied to study for a teaching certificate and looked for a school that emphasized attention to educational problems. This became my challenge, and I believed that with faith, love, and devotion, I could make a difference to society.

It should have come as no surprise that there were many such schools waiting for me to apply, since teachers usually struggle to find a place in "good" schools and shy away from the "problematic" ones. As a result, the latter institutions are usually staffed by idealists, alongside teachers who did not take the trouble to look elsewhere or simply had no alternative.

This school (which actually called itself an "education center") was situated in ironic proximity to Abrabanel psychiatric hospital in Bat Yam. I was warned that most teachers "leave after two weeks," but the two years I spent there were full of wonderful experiences. There were several good teachers there, and I learned a lot from them. I noted that some 97 percent of the students at the school were of Mizrachi origin, and I came to believe that many of them came to the institution after being denied proper opportunity at their previous schools. Some teachers, however, pointed an accusing finger at the students' parents.

I decided to devote all my attention to my students: to teach the various subjects at the highest possible standard, to give them warmth, to meet their requests, and, above all, to be open to them. I visited many of them at home. The students returned my affection, although I was sometimes unsure as to whether I could accept their offerings. Once, for example, after I mentioned that I did not have a radio-tape in my old car, they appeared next day with a radio. I would not agree to accept the gift and was worried about how they had acquired it, but I thanked them warmly for their thoughtfulness.

To this day, I sometimes meet one of my former students by chance in Tel Aviv. It always give me pleasure to meet a shopkeeper or a small businessman and find that they are proud of what they have achieved. Above all, I am pleased to learn that their own children do not go to the "education center." I am not sure how significant an influence I had on them, but they claim that it made a difference.

Around the same period, the Jewish Agency was looking for people to work as community emissaries abroad. I went to the interview casually, certain that I would not be accepted (after all, most of my educational experiences had given me the message that there was something wrong with me). After an eighteen-month selection process, we found ourselves—as a young couple—living in Tulsa, Oklahoma. Tulsa has a small, affluent, Jewish community of some 3,000 souls with a highly developed communal life. As two secular Israeli Jews, we suddenly found ourselves living in a small Jewish minority in the middle of the fundamentalist Christians of the mid-West. We went to all the synagogues and served as representatives of all the Jewish streams, but we found ourselves particularly envious of the liberal streams, which presented a Judaism that was noncoercive and attractive—quite different from what we were familiar with in Israel.

My work as an emissary required me to work with people of different faiths, a task I undertook with enthusiasm. I had several interesting meetings with members of Muslim community in the city.

After three years as an emissary, I stayed in the United States for two more years in order to study for my MA degree in Jewish Studies. I studied at JTS (Jewish Theological Seminary) in New York, the school of higher education and center of the American Conservative movement. My MA studies eventually led to doctorate studies at the same institution.

Education studies at JTS are combined with theoretical studies at the Teachers College of Columbia University. I studied qualitative research methods and statistical analysis. I also became familiar with the anthropological approach to educational research, and with other aspects that were useful in preparing the present study.

My doctorate thesis related to the development of the identity of young Reform Jews in the southern United States, including an examination of the consequences of their encounter with Christian fundamentalism.

After we returned to Israel, I worked for two years in Melitz, an organization established by Avraham Infeld that uses advanced counseling techniques to examine questions of Jewish identity. Part of my work was being responsible for the international management of the "Pikudei 2000" program, which encourages independent research into the Jewish family by children of Bar/Bat Mitzvah age. During the same period, I also began to teach at the David Yellin College of Education. After two years of educational work at Melitz, I began to feel guilty that I was not giving anything to Israeli society. I looked for a heterogeneous framework that would enable me to return to direct teaching. I found such a school in a community near Jerusalem, a place that had came under fire several times during the Al-Aqsa Intifada.

Unlike my first experience working with "difficult" students, this time I came to the classroom after I had acquired an intellectual foundation from Kozol, Holt, Freire, Buber, Korczak, Neill, and other good friends.

I believe that my experience with my new students was no less powerful than my first experience in the field, although the feedback I received was mainly from the students and parents. I did not feel that the school had any forum encouraging the staff to reflect on their work, and so no real discourse took place.

I do not think that there was anything wrong with the teachers at the school. Rather, the system that had developed there failed to encourage real communication based on trust and a need to share both successes and failures, problems and solutions, with one's peers, in the knowledge that any human endeavor, education included, will always include an element of the unknown and the unfathomable. As educators and education researchers, all we can do is engage in interpretation and dialogue in an effort to understand educational processes, while simultaneously creating positive conditions for the continued development of a positive educational and academic experience.

By the following year, I was already working full time at David Yellin College, where I continue to work to this day. The seminary includes many Arab students and lecturers, and I have had an opportunity to work to advance coexistence between the two peoples and to discuss issues of multiculturalism.

A forum of Arab and Jewish teachers has been established at the college, it meets once every three weeks for a dynamic workshop that attempts to examine different aspects of coexistence at the seminary, in Israel, and in the region as a whole. The sessions are difficult and sometimes tense, but I believe that all the participants gain from this meaningful encounter with the other, and the exploration of the possibilities for cooperating despite the differences.

Together with other teachers at the college, I have been able to develop study modules designed to encourage teachers to examine their own work from a more critical perspective, and thus help enable them to make changes and improvements based on power and knowledge. We have encouraged those working in the field to undertake practical research, usually based on qualitative research tools, although teachers at the college have often been able to assist necessarily with quantitative research tools as necessary. These modules aim to fill a vacuum that I have encountered in several frameworks over the years. By contrast, in several of the alternative schools I visited, I found working methods that seem to constitute a form of ongoing collaborative research. This is particularly true by comparison to more traditional and centralized schools.

We are increasingly reaching the conclusion that the most important training for those who work in the field takes place in the field itself. As part of the final seminar for the fourth year students, we took them to visit the schools described in this study. The students' reactions provided additional and interesting data for the research.

I have been a minority in another country, an immigrant in Israel, an emissary in the Diaspora, and an activist and researcher in various frameworks for intercultural encounters. In all of these situations, I have encountered the other. I have no doubt that these encounters have heightened my sensitivity to minorities and to those who are different, and predisposed me to be sympathetic to those teachers who have made Repair the formative ideology behind their educational approach. My encounter with the schools that operate on the basis of the Ideology of Repair is also my encounter with myself.

CHAPTER ONE

Neve Shalom / Wahat Al-Salam

(Grades 1st–6th, approximately 260 students)

When two populations are embroiled in a protracted and existential conflict, clinging to myths that offend or negate the other side, the usual assumption is that each population will wait for a gradual thawing of relations creating the conditions for mutual understanding and cooperation. The residents of Neve Shalom did not wait for "the processes to take their course" or for "confidence-building steps." They addressed the difficult task of living through the conflict by establishing a joint village for Arabs and Jews. The school opened in the village and originally served the children of local families, but later also absorbed many students from far and wide—from Mevasseret Zion, Ramle, Lod, Zur Hadassah, Abu Ghosh, and elsewhere.

The school in Neve Shalom is sometimes confused with the Peace School that is also based in the village. The local school provides formal education to children from the village and the surrounding areas, whereas the Peace School concentrates on providing services for different groups through workshops and seminars. The two institutions function alongside each other, but have distinct ideologies and methods. There is no formal or institutional relation between these entities, although the residents of Neve Shalom compare their approaches. Over the years, considerable changes have been made at the school. It has expanded greatly and is now preparing to receive full recognition from the Ministry of Education.

The school is managed by two principals who are selected for a three-year period that may be renewed. The current principals are Boaz and

Diana. Boaz is responsible for "foreign affairs" (such as contacts with the authorities and the Ministry of Education), while Diana concentrates on developing and supervising the didactic and academic work of the school. Principals who complete their term of office are offered the option of remaining at the school as teachers, consultants, or in other functions.

Diary[1]

I arrive to visit Boaz and Diana. The village is situated on a hill with an impressive view, not far from the site of one of the battles of the War of Liberation, or the Naqba (depending on one's perspective).

The village is growing steadily, and the residents seem to be gaining confidence in their presence on the ground, although fears are sometimes evident. Boaz used to be responsible for security in the village (he is a former IDF officer). Once he and some friends came across a group of people walking around the area. The visitors told him that they were members of the Special Operations Unit who were due to move into homes in the planned new neighborhood of Neve Shalom—a plan that was prepared without the knowledge of the secretary of Neve Shalom. The residents fought against the plan. It was eventually shelved, but there have recently been signs that it may be proposed again.

I arrive at three o'clock in the afternoon, as the school day is drawing to a close. I am met by a blur of sound in Arabic and Hebrew. Drivers shout to the children to get into one of the buses that will take them home. Yaron, the teacher responsible for overseeing the transport arrangements, uses a megaphone and tries to impose order.

The school building is reminiscent of a kibbutz school. Indeed, Neve Shalom as a whole has something of the appearance and structure of a kibbutz. However, when the village was still young, its members decided not to use the term "kibbutz," with the negative Zionist connotations this carries for some of the residents, to describe this Jewish-Arab village. Many residents see the kibbutz as a place that denies the basic equality of the two peoples. Accordingly, they chose a more neutral and acceptable term: village.

The school occupies several buildings around a large and attractive plaza. One wall bears a picture of a rainbow with the name of the school in both languages. I would often sit here and watch the interaction between the Jewish and Arab children during recess. Boaz welcomes me to his office. He is always dressed in jeans and checked shirts, just as many of the members of my kibbutz used to wear in the past. Boaz grew up on Kibbutz Shoval. Some of the parents, mainly Arabs, feel that his style of dress is not really suitable for a principal.

Diary—In Diana's Home

On one of my visits to the school, Diana invited me to her home, which is situated just a few meters from one of the side entrances to the school. We sat and looked at the view together.

Diana

From our house, you can see Tel Aviv and Lod (she is silent for a moment, and then shifts her glance) . . . And over there is occupied land (another pause). It is no coincidence that the village was built close to no man's land. We were the third family to come here, and the first Arab family to build our home in the village. We didn't have any money, but our friends came here to work with us. They didn't always consult us, so once again we learned that everything has its price. We wanted an Arab-style house, with a broad entrance and fountain. The roof is reached by stairs from the side, and it must be flat so that you can lay out fruit and vegetables to dry. Over here (she points) there is a pair of small, narrow windows that offer relief from the long summer. Diana notes that this pair of windows is typical of Arab homes. She points to the grapevine outside the window, and adds: "A vine outside the home is a very important symbol."

The house combines Western and Oriental elements, creating a pleasant harmony. Diana points to a large window in the dining area. "We should have opened up a window to the view here. We couldn't do everything the way we wanted—there were financial constraints."

Our time was limited. After a brief chat, we went back to the school. The rapid transition from the school to Diana's home and back helped me appreciate the profound bond—even in simple physical terms—between the village and the school. This proximity also emphasizes the complexities involved in running a school based in a village with a clear political and ideological orientation, but which is home to teachers and students whose attachment to the village and to its ideology is often limited.

Equality and the Internal Process in the Intercultural Encounter

Of all the intercultural encounters created within Israeli society, none would seem to be more charged than the one between Jewish and Arab citizens of the state. The school in Neve Shalom thus faces the greatest educational challenge of all the institutions discussed in this study.

As a generalization, intercultural encounters are unsuccessful, because humans tend to adhere to an ethnocentric approach that protects them against the threat of exposure to the other (Pearce, 1982; Brislin and Yoshida, 1993). Virtually every culture in the world is exposed to other cultures, and each intercultural encounter challenges all those involved.

Tourism is a popular channel creating intercultural encounters, though these generally tend to reinforce ethnocentric positions. Limited exposure to the other culture strengthen existing stereotypes regarding the host culture, particularly through the exchange of impressions with other tourists in the group (Pearce, 1982).

Many Israeli Jews have few encounters with Arab society and culture, relative even to other, more distant cultures. The encounter between the two cultures is one of dominator and dominated.[2] Human friendships are usually created when a state of equality exists between the two sides concerned. The encounter between Jews and Arabs takes place in the context of employer–employee relations. Lasting and respectful relations may develop in this context, but there is a silent agreement to avoid discussing certain subjects, and the relationships do not usually develop into a real friendship.[3]

Situations of equality between Jews and Arabs are rare. When such situations are created for limited periods—through encounter programs for youths or students, or in other frameworks—they engender a sense of embarrassment and fear, particularly among the Jewish participants:

> Yes, I have met Arabs before, but not like this. I met my grandfather's workers. But it was different with them, and here I realized that . . . well, I didn't really like the idea of sleeping in the same place—in the very same tent.[4]

The school in Neve Shalom attempts to create an equal encounter within an unequal society. The aspiration to equality is evident on all levels and on an ongoing basis. As early as 1969, Amir[5] rejected the argument that the mere act of the intercultural encounter would improve attitudes toward the other side. Amir proposed a list of conditions that are required for a successful encounter—conditions that the school at Neve Shalom attempts to create:

1. The attitude to all the participants must be egalitarian.
2. The encounter should take place between members of the majority group and members of high socioeconomic status in the minority group.
3. The positive atmosphere should encourage a positive encounter.
4. The encounter must be more intimate and less group-based.
5. The encounter must be enjoyable.
6. The participants must together engage in meaningful activities and must aim to achieve similar goals.

In addition to the ongoing attempt to create an egalitarian encounter within an unequal society, and to the effort to maintain equality on all levels, the school community in Neve Shalom also aims to encourage all participants to undergo an internal process. Ideally, the encounter with oneself and with the other, and the complex acquaintance with the other, ought to lead the participant to encounter internal dimensions of otherness.

Diana recalls the different stages she went through before reaching the current stage. She feels more at peace with herself now, although she believes that she will always continue to develop in new directions:

> In the 1970s, I began to feel hate—something I didn't want to feel. I had my reasons for hating. At school, they didn't teach us anything about our culture or identity. But I bought books and studied by myself. At university, I began to hate the other side for stealing our dream. Later, after participating in workshops, I realized that at the end of the day I have to try to help myself and examine the way I see things, how I feel and how I analyze situations. I can't just take things for granted. Later still I attended Reiki workshops and other alternative processes.
>
> I have to look for the light in myself and in the universe. That strengthens me and gives me the energy to continue.
>
> Today I like the place where I am. My work is not easy, and it has got even harder recently. But we can be proud at this school, despite all the problems. I think that anyone who makes peace with themselves can make peace with others. Indeed, if I want to reach out my hand in peace I must first examine my own position in the

> whole story. If there are barriers here, why do they exist? What do I think of you as a Jew? Can I agree with what you say or not? I analyze my answers to these questions.

Here, Diana reveals biographical details of the process she has undergone. She emphasizes the importance of introspection within the context of encounter with the other. Not all the teachers, parents, and students at the school are engaged in such a profound process, but each principal provides a role model for coping with the conflict while engaging in introspection to resolve conflicts within one's own self. Diana assumes that once a child has developed a confident identity, understands himself, and accepts himself and his culture, he will also be able to accept and understand the other without feeling that his own being or existence are threatened. In this way, those involved in the process come to see the other as differently cultured but equally human.

> Boaz
>
> We are involved in the search for identity and in developing our identities. People believe that when someone has a clear perception of their identity, they will be less threatened by other identities. The same applies when people use words with national connotations, such as Arab or Jew or proud Palestinian. We don't avoid the word "Arab" here. A comment such as "the Arab children are on vacation today" does not raise any problems for us. It's part of life, just like saying "boys to the right, girls to the left." That's how we learn to deal with things. It's part of our identity, without any great trauma. Respect, acceptance and the place given to the national identity of each child and of the other enable the children to grow up confident and proud of their own identity.
>
> In Jewish society, the word "Arab" is a curse. Not here. At our school the children have to look for other, more effective curses.
>
> When the school was established, it was intended to meet the needs of the children from the village. We emphasized the three religions and the festivals of each national culture. We don't talk about it—we live it. It isn't a matter of "you should be nice to . . . " or "treat equally"—but a matter of living together. As someone who was born on a kibbutz, I am aware that when you speak about equality you raise expectations, and this can sometimes lead to disillusion. It's a fascinating dilemma. But Neve Shalom is a place where we have really been trying to maintain equality for a long time.

The internal process of the individual and the group in Neve Shalom is based on the basic conditions of equality at all levels of the encounter.[6] As we have already noted, this is an equal encounter within an unequal society. As will be discussed below, the Arab side is usually particularly sensitive to possible signs of inequality that infiltrate from the surrounding culture and the broader context of Israeli society. What happens in Neve Shalom is not a sanitized effort to be nice to one another. The school community allows its members to live with the conflict, with all its complexity. The participants are invited to experience mutual and parallel processes through which they gain familiarity both with the other and with themselves.

Changing Attitudes and Expanding the Borders of Responsibility

As in the example noted above—"The Arabs are on vacation today"—words undergo a semantic transformation at Neve Shalom as a result of the internal process undergone by the participants. In Neve Shalom, the word "Arab" and all its derivatives become a neutral and/or positive concept. From the perspective of an observer from an outside society, the Arab or the Jew becomes part of a human community toward which the child feels affection, identification, and concern. The changing attitudes among the children sometimes surprise the parents, who also appear to be influenced by this process.

Diary

A group of parents is interested in sending its children to the school at Neve Shalom; some of them already have children at the school. One mother relates:

> When the Intifada broke out, my other son's class—at a regular school—also discussed what was happening, but in the form of explanations: what is the Temple Mount, why is it important, and so on. But my daughter, who goes to the school at Neve Shalom (3rd grade) came home worried at the possibility that war would break out. I tried to reassure her: "We are strong, even if there is a war we are strong and we will win, it'll all be ok," etc. But my daughter said, "Mom, you don't get it. No-one wins in wars.

People get killed. And if we win, who will we kill? They are my friends, my relatives—what will happen?"

Boaz believes that this story is typical of the process that occurs at the school. "This girl is not quoting someone else. These are values that she has developed at the school." The school indeed seems to contribute to the development of identities alternative to one that the mainstream of Israeli-Jewish society offers. Jewish and Arab children learn to think about the results of the conflict from a broad perspective, and this thinking process becomes more profound as more instances of conflict occur in the external environment of the school. It is no longer a matter of "them and us." The problems together become a concrete threat to people with whom the children live every day.

Changes can be seen in everyday happenings. A mother who went to a shopping mall with her daughter saw her daughter stop after she saw an Arab family at the entrance. The daughter explained that she wanted to see how the family would be treated. Would they dare talk in Arabic? What would happen? How would they be treated?

Boaz

The girl was showing a level of sensitivity that surprised her mother. She hadn't expected that. The child is sensitive to details that are ignored by adults. No-one told her to watch what happened in her surroundings. The girl had internalized the issues.

Diana often talks of the changes that have occurred in the children, noting that these occur much more easily than processes of change among adults. Adults must overcome an ancient archeological structure of barriers and prejudices. The desire to break these barriers eventually led Diana to Neve Shalom. Even today she is delighted when she sees the alternative socialization undergone by the children in the community.

Diana

In 1984, we moved to the community as a family. Our two children were born here. They are lucky to have been born in a place like this. I am learning through them and through the children at the school. They don't have the barriers that we adults have. We build barriers for ourselves, and then we suddenly notice them and work hard on ourselves to break them down so that we can see the other side from a perspective of equality, appreciation and even a little love. I really see these things through the children.

For the children, it is all very natural. They can sit here in the house and shout about some political event, raise opposing views and then continue to be friends, and that is fine. Over the past year, some very difficult events have occurred in Israel. But that has not prevented my little son, in the sixth grade, from inviting Jewish friends to sleep over, or going over to their homes.

The children organized a party and the parents called us and said how great it is that we have the energy to meet even in the current situation. I have always said that we need to learn from the children, not vice versa. That doesn't mean that they do not have a political conscience—they are actually very involved politically and understand what is going up. But they have the ability and the willpower to keep on living and to enjoy the brief lives that we are all given.

Diana Adds

Sometimes five or six friends of our children visit the house. When they come, they do not think "Now I'm going to an Arab home or a Jewish home." It is also natural and automatic. Since the 1970s I have had to work so hard on myself in order to realize that the Jew is not an enemy, period, but a person who thinks and feels and believes in things—sometimes contrary to what I believe in, but I accept the complexities of my own people and of their people. Everyone has their own ideas and thoughts. If I accept myself, why shouldn't I accept them? Everyone in this country thinks that their side is OK, their thoughts and actions are legitimate, but those of the other side are unacceptable.

Diana on the Children

The children here live the Arab-Jewish conflict. How can you cope with this complexity in a way that turns it into an advantage? Take the example of the vacations. One girl postponed her birthday party because the Arabs were on the winter vacation, although her mother wanted to go ahead with the party. Another girl said, "Why don't we have the party in class after the Jews come back?" For the children, both sides are positive.

Diana and Boaz often observe, document, and engage in reflection relating to the behavior of the children in the peace laboratory in which they live. These stories are created in the community and in the school, and there is an evident connection between the two that is not free of problems. Through the children, it becomes apparent just how artificial and dissolvable the archeological strata of hatred and barriers are. The community and the school engage in socialization for peace and conflict resolution. Each transforms the other into a living, real person. When the girl cancelled her birthday party because her mother would not allow her to invite her Arab friends, she emphasized the difference in attitudes. For her mother, "Arab" was an abstract, distant concept. The daughter, meanwhile, thought of the presence of all her Arab friends as being so meaningful that it would be pointless to hold the party without them. Clearly, the school makes a real contribution to interpersonal rapprochement, understanding, and acceptance.

In educational terms, this is an example of prefigurative learning.[7] The children cannot learn proper models for conflict-solving behavior from the adults. In fact, the adults learn through the children that there is a possibility for a better future.

Socialization for Change

The school makes an effort to create change among the children, the teachers, and even the families involved in the school community. In some cases, the children become agents of change in their family, concretizing themes that the adults have ignored. Among Israelis who define themselves as left-wing and supporters of the peace process, many have no real acquaintance with Arabs. The school facilitates a process of socialization, familiarization, and acceptance, making them aware of the inadequacies of their perception of the other in the context of the Arab-Israeli conflict. Equally, however, it is obvious that the parents of the children have an underlying ideological support, at least to a certain degree, for the process implemented in the school. Many families from the mainstream of Israeli-Jewish society would almost certainly react with apathy, neutrality, ambivalence, or even hostility to such manifestations of "insights" among their children. As a generalization, people who are prone to suffer from culture shock are actually less ethnocentric. Ethnocentric individuals completely negate the other culture and judge it solely according to the standards of their own culture; accordingly, the inherent veracity of their own culture is not challenged. Nonethnocentric people are more willing to examine the "truth" of the other group; when their preexisting understanding of reality is inconsonant with the new context, culture shock is created (Brislin and Yoshida, 1993; Pedersen, 1994).

The Parents' Motives

Although the school in Neve Shalom is only partially associated with the community in which it is situated, the cumulative experience of shared life influences the spirit and perception of the school. However, the school also serves a large number of children who do not live in the community. What motivates parents to send their children to a school whose perspective is not always consonant with their own? The answer seems to lie not in a desire to participate in a bicultural, Arab-Jewish project, but rather in a desire to send their children to a high-quality, relatively small, intimate school with a good academic level. All these features can be found at the Neve Shalom school.

Many parents, Jews and Arabs alike, state that the high standards at the school are a very important factor in their decision to send their children to Neve Shalom—a decision that sometimes entails long commutes to and from the school.

Among the Arab parents, I found a higher level of concern regarding the questions of excellence and standards. They want their children to acquire appropriate tools for coping in a society that discriminates against them, creating a "glass ceiling" that blocks opportunities for advancement. Despite these restrictions, the parents view education as an important tool for upward mobility in Israeli society. The Jewish parents seem to be more open to an alternative and less demanding educational style, although here too, voices may be heard that seek "excellence," whether in terms of grades or in terms of input and output.

Diana is critical of the prevalent position among many Arab parents (including families who live in Neve Shalom), and also among not too few Jewish parents, though she understands the source of their concern. She appreciates that other people's experiences are different from hers and accepts that many parents have a different educational approach.

Diana

I wouldn't want my children to be educated the way I was. I believe in a multi-capability approach in schools. We have to give children different opportunities, tools and capabilities, so that they can move on and make progress. Philosophers and educators have spoken of a new age in which things are changing. Carl Rogers also said that we must constantly return to ourselves in order to survive. As teachers, we need to understand the fact that the children are always rushing forward. We need to give them a bit of support, and say, "Go ahead! We'll run with you." That's very important.

We do not always succeed. It is difficult for us when a parent doesn't understand why we do not emphasize achievement and tests. The difficulty is particularly apparent when a new teacher comes to the school and starts giving grades and examinations, or when the teacher gives in to pressure from the parents.

If you give a grade of 100, so what? And if you give a grade below 80, you cause damage. What you need to give is evaluation that helps the child move forward. A grade just creates a barrier in the child's way. This is what happens at the Orthodox school[8] and elsewhere. The child needs to say: "I need to study so that I can understand more and more things, not for examinations or grades."

It is important to recognize diversity. That reminds us of the big rainbow in the school. Everyone has their own rainbow with many colors. We only need to let the child find out what they need to improve.

If we had a democratic school here I would be happy and support it. At this stage, a democratic school is not appropriate in terms of the dynamics in Neve Shalom. People want the security that comes with an emphasis on academic achievement, and they aren't yet fully satisfied or confident about their children's achievements. This problem is more evident among Arab parents than Jews, but Jewish parents are also sometimes concerned about achievements. I find this need surprising. The Jewish parents say that they leave their children in the school because the children enjoy it, but sometimes they have real doubts whether it wouldn't be better to move them, because they are not learning enough, or they aren't learning "the way they should." When we went to look at the democratic model in Hadera, we had a course and we tried to implement a particular direction and some parts of this approach. I have thought a lot about democratic education as an alternative to the emphasis on achievements, but I realized that if the parents cannot identify with this educational approach, it won't work. Children need support from home, and if I come from a more traditional and authoritarian society, it will be hard for me to support the democratic alternative.[9] If I could, I would build a democratic school here in Neve Shalom along the lines of the school in Hadera, but with adaptations to meet our local needs. But as in

> Hadera, the basis has to be that the child chooses. I believe in the abilities of children, they can learn when they want to and overcome any gaps in a very short and intensive period, with the right guidance. Why not? When people said that the children here just play, I replied that children learn by playing, too, and later they integrate socially into the schools they move on to. When our children finish the 6th grade here, they stand out wherever they go to study, so why should we pressure them? Most of my time at school, when I was a student, I felt pressured.
>
> We need to go through a correcting experience through our children, not to make them go through something that we didn't like.

The positions of Diana and Boaz regarding the educational process, like those of the principals we met in the other schools, have been influenced by the theories of humanistic psychology (Rogers, 1961, 1973; Maslow, 1964, 1968), by theories relating to different learning styles and by multiple intelligence theories (Dunn & Griggs, 1988; Gardner, 1992). The school emphasizes the importance of nurturing emotional intelligence as expounded by Goleman (1997) and attempts to meet the need for education for life in an age of constant change.[10] The democratic school in Hadera provides the school in Neve Shalom and Diana in particular with a useful model for inspection and examination.

As in some of the other schools (as we see below), we see here a phenomenon whereby the educational leadership is at an ideological and philosophical point that is different from the parent, in general, the leadership is more radical, more democratic, and more alternative. The desires of the parents and that of the educational leadership are not completely consonant with one another. Accordingly, the school ethos is influenced by the mediation between sometimes contradictory trends among the educational leadership and the parents.

The same is true of the teachers. The principals have difficulty in convincing the teaching staff of every aspect of their approach and educational worldview, but many teachers are however influenced by the principals' leading discourse.

The teachers in the school often commented that there are many "good" students at the school. I spoke to Boaz in an effort to clarify the covert meaning of this term "good." I asked him whether he is not concerned that only a certain kind of student is being "directed" to the school. Is it possible that the school actually offers an encounter between elites from both sides?

Boaz

In the case of the Jews, there is a bias. There are not many Mizrachi students and not many weak students. Most of the families are middle-class Ashkenazis from prosperous communities, although there are some families with children who have difficulties. The Arabs are more diverse. This is a good school, and you may be right that it attracts Arabs who want to enjoy the standard of learning of Jews. For some of them, the school is a very important lever for improvement. But we also have students from very deprived neighborhoods of Lod and Ramle, neighborhoods that have serious problems of drugs and violence, such as the "Train Neighborhood," Jawarish and some neighborhoods that don't even have a name. One of the children at the school walks two kilometers from the drop-off point to his home, because the driver refuses to enter the "drugs neighborhood." I asked the driver why an eight-year old boy has to walk so far: "Why can't you take him to his home?" The driver told me: "I have had stones thrown at me there, they broke my window and I'm not willing to go in there. Apart from that, every time I go into that neighborhood the police stop me and search my car." The father of one of the children at the school was murdered in the city, probably in a crime-related incident.

A problem emerges of a discrepancy between the students and the social classes. This is also true of the Arabs from higher socioeconomic backgrounds. Among the Arabs, there is a desire to avoid inequality, but they also want their children to achieve excellence, and the encounter with the working class is problematic. For the Arabs, the school may tend to be slightly elitist. The parents are looking for discipline and homework; they want to see that their children are learning a lot here. They don't want their children's future to be blocked. But we feel that we are doing a god job even if we don't give lots of homework and tests. We see that our students settle in well in the junior high schools. For example, the Orthodox School takes the children from Neve Shalom, but they also take our children from Jawarish. The discrepancy also applies regarding the expectation of strict norms of behavior and achievements. Money does not present a barrier for the children from the disadvantaged neighborhoods. It's true that the parents of children here pay higher fees than at regular schools, but help is available for those who have problems. The children who come from the neighborhoods are generally a little more aggressive, but they change during their studies. They become more moderate. I see that as success.

> One mother told me: "I have five children who all came out the same, but Mohammed, who went to your school, isn't like the rest, he behaves completely differently." She is sorry that he will have to go back to a school in his city when he finishes the 6th grade. In general, the perception among the parents is that the children at Neve Shalom are different . . . We don't see it so much here on a daily basis, but the parents see the differences.

> Diana
> Some of the Jewish students manifest the problems of "modernity"—divorced parents, parents who are absorbed in their careers, etc. These problems can be seen in the daily behavior of the children in school.

Thus we see that the desire for excellence is very important to both populations. The school in Neve Shalom follows Amir's recommendation[11] that an encounter will be more successful if the participants belong to similar classes. This principle is followed, even though, as noted earlier, several Arab children come from areas that are considered "difficult." From an observation of the children's behavior, it seems that the school is successful in achieving positive socialization. These children tend to become less aggressive and manage to control their feelings better, which makes them stand out in their home environment, as noted by parents who sent some of their children to Neve Shalom and others elsewhere. In other words, the students who come from the problematic neighborhoods do not generally change the school; on the contrary, they themselves are changed by the school society. One of the problems is that in these neighborhoods there are no junior high schools that can provide a continuation of the positive experience at Neve Shalom. Therefore, the school is currently working to establish its own junior high school, with tentative plans for a high school. It should be noted that among the Arab teachers and parents, there are those who favor a reduction in the proportion of Arab children from difficult socioeconomic backgrounds. This segregationist approach is also found among the prosperous Jewish population, which tends to classify itself according to codes of social, cultural, and economic status.

The parents of the Jewish students at the school are mainly educated, Ashkenazi, and well-off. Many of them belong to the elite that still controls the power foci of Israeli society. It is a well-known paradox that the Jews of Arab descent, who have an affinity to Arab culture, often attempt to underplay their identification with Arab culture and with Arabs in general.

Moreover, the Mizrachi Jews are generally more traditional in religious terms, and Neve Shalom does not seem to devote significant attention to the study of the Jewish tradition.

> Diary
>
> I look around the outside of the pre-school and then walk inside. The pre-school teacher, a Jewish immigrant from Russia, is explaining to the students how the children learn about the three festivals—Christmas, Ramadan and Hanukkah, which this year fall very close together. I notice that there is no mezuzah on the door. Are there any mezuzot in the school? On a symbolic level, it occurs to me that the school is not a welcoming place for some of the Jews. I saw a student who is used to kissing the mezuzah look at the doorpost in amazement, as if she were lost. This is another result of the intercultural encounter.

Despite the self-selection, despite the fact that the populations at the school are educated toward the idea of encounter, unexpected difficulties are still encountered. The school in Neve Shalom reflects neither Jewish nor Arab society as a whole, but it can provide lessons as a pioneer in meaningful experiment in coexistence between two communities that are in conflict and seeking identity.

Myths in Conflict

A myth is the raison d'être of a particular community. A myth is a story that shapes the culture of a community and explains to its members who they are, where they are heading as a group, and what distinguishes them from any other communities (Gillman, 1990).

Sometimes the formative myth of an individual is in conflict with the dominant myth of the surrounding society. In the United States, I encountered a type of reactive and negative Jewish identity among youngsters who lived as a minority within a fundamentalist majority (Bar-Shalom, 2001). The youngsters were disturbed by the fact that their Christian friends were sorry that the Jews would burn in hell because they refused to accept what from their perspective is the only correct myth. There is a widespread assumption that the current global wave of terror has religious causes. Extremist Muslim fundamentalism advocates a holy war to liberate the entire world from the infidels, who must either die or convert to Islam. Christian fundamentalism awaits the coming of Christ, and a preliminary stage in this process (according to pro-Zionist Christians) is the return of all the Jews to the Holy Land. Fundamentalist Jewish extremists seek to blow up the Al-Aqsa Mosque in order to expedite the coming of Messiah and the establishment of the Third Temple, reflecting their belief that only Jews are worthy to live in the Holy Land.

These are examples of powerful myths that dictate an ethos of conflict, ethnocentrism, and the complete negation of the other. How do cultures that negate each other's existence operate together and cope with the conflicting myths?

As if the coexistence of three religions in Neve Shalom School were not enough, there is also a national conflict here that also includes conflicting myths: the myth of victory and liberation in the War of Independence versus the Naqba,[12] a myth of destruction and victimhood. The same historical event is interpreted in opposite ways depending on the national affiliation of the observer—a telling example of the social construction of reality.

In Neve Shalom School, it seems that it is not inevitable that mutually exclusive ethnocentric myths and ethoses will clash. Although all three religions include engrained myths and ethos systems that have led, and still lead, to terrible conflicts, there is also some evidence of alternative, "softer" myths that offer the individual community, belonging, and symbols without the need to negate the other. It is beyond the purview of this book to examine the different streams that may be encountered, but we can note that within each of the three religions, the canonical texts permit an interpretative approach that influences the development of different types of worldviews. Within each of these religions, there is a tension between two streams: streams that negate contact with and study of others and streams that encourage self-examination through encounters with the other, fostering tolerance and seeking for themes that unite all religions and cultures.

Along the cultural and religious continuum of festivals, customs, legends, and texts, Neve Shalom appears to be seeking the "soft" myth that enables friendly and warm contacts between the different communities. The children speak excitedly about the upcoming festivals. Sometimes the festivals of both peoples fall almost simultaneously, and a decision is made as to which festival to celebrate. Sometimes one is concerned that because of this constant exposure to so many traditions, the children may grow up with a confused sense of identity.

Boaz (Discussing His Son Tom)

At first I was worried what I was doing to the boy. What would happen when he grows up? First there was the Hijra, when they celebrate Mohammed's journey to Mecca. Then there was Christmas, and then Passover and Moses. I was afraid that they might be damaging the child's identity. The children are preoccupied with the festivals. They all get excited about who is going to be Father Christmas and bring the presents (one of the teachers dresses up for the occasion). I asked the children if all this doesn't confuse them—Christian festivals, Muslim festivals, Jewish festivals. One of the children told me: "When I go to a friend's birthday party, don't you think I get excited? But I know that it's not my birthday. So what's the difference? Why should it be any different?"

This is an example of how adults learn from children. In Neve Shalom School, and in most of the other schools we visited, the children have many insights that influence the understanding of the adult teachers. Much of the socialization takes place outside the confines of the school, so that the exposure in the school to other cultures does not threaten the children's own identity. Naturally, they get particularly excited about Father Christmas coming to visit, because he brings presents—and who does not like to get presents? But the children understand and sense that the fact of participating in another culture does not weaken their own identity. The opposite may be true. Ruth Gabison has claimed[13] that some parents send their children to binational schools such as Neve Shalom precisely because such schools require the students to confront their own Jewish identity. Similarly, we hear that many Israeli backpackers discover their Jewish identity after traveling through the Far East. This must be qualified, however, by noting that the encounter sometimes does encourage the adoption of customs of the host society.

The myths relating to religion and tradition may be resolved to a certain extent. We do not know whether such a dynamic could have emerged if those involved in Neve Shalom came from more traditional backgrounds. It may be assumed that participants who come from a universalistic and diffused background will find it much easier to adopt the customs of others than those from particularistic and iconographic backgrounds; as noted, however, the latter constitute a minority in Neve Shalom.

National myths are harder to resolve. These myths still create difficulties within the broader Jewish and Arab society in the region, and within the microcosm of Neve Shalom.

Independence Day and the Naqba are marked separately in the community and in the school. Some of the residents of Neve Shalom, particularly Arabs, would like to bring the whole community together to commemorate the Naqba and to avoid marking Independence Day (based on their belief in the importance of the Arab sense of loss and the need to mark this aspect), which most of the Jews in the community identify with. A paradigmatic inversion may be found in Neve Shalom, from a situation where the Arabs living in Israel are unrepresented, absent, and lacking a voice and influence to a situation where they are represented, influential, and having a voice or narrative that is heard and present. The Arabs in Neve Shalom have a position, power, and influence. However, they sometimes seem to interpret Jews' disagreement over a particular point as a return to the pattern of silencing the Arabs, oppression, and the like.

Boaz

> The narrative divides the members of Neve Shalom. The narrative of most of the Jews is different from the narrative of the Arabs. We don't have one answer. Some people in the village find this complexity difficult. For example, they want us just to mark the Naqba—why should we split up on this day? Those who are looking for exclusivist positions will usually find them on the Palestinian side, because they still act as an engaged society, as we were during the War of Independence. Engagement was needed then in order to achieve the goal. Some Jews give up their ethos and history because of the difficulty that comes from the fact that someone else views these differently. A Jew can move toward the Arabs, but a Palestinian will find it hard to do the opposite. The Palestinians are aware that they are still fighting for basic rights, such as their own state and equality. It is difficult to manage with two parallel stories. We are often in a situation of "versuses" here—one thing versus another. This isn't easy, and those who initiated this idea and established the community may not have been aware of the difficulties. In the distant past, in conflicts within Israel, the Arabs always gave up on the important aspects. Today, the Jews are more inclined to give way. In school, in discussions among the teachers, there is a sense that the narratives are drawing closer.

Within the conflict, Boaz has undergone the hardest experience of all: the loss of a child. His son Tom was killed in a major military helicopter accident on his way to Lebanon. The media reported prominently (and inaccurately) on Tom's funeral, an event that led to tension in Neve Shalom. Some of the Arab members of the village mistakenly got the impression that Boaz wanted to hold a "military funeral" for Tom, or to erect a memorial for him. Disinformation was rife, and Boaz believes that there were those who found it difficult to see him as a victim, because the myth of Neve Shalom generally identifies victimhood with the other side. A compromise was eventually found, but Boaz's children still have some resentment. Boaz explains: "When you are in a position of power, it is easier for you to give ground. Some of the Arab members found it hard to give ground because of their 'sense of weakness'." Boaz's philosophical insight does not seem to have convinced his children.

Boaz

On the day before Memorial Day for the fallen soldiers, a problem arose. I was invited to light the torch at the main ceremony on Mount Herzl at the beginning of Independence Day. Some of the Arab members asked me how I could possibly agree to do this. After all, people would assume that I was acting on behalf of the whole village. There may also have been a question of honor—due to lack of knowledge, they felt that I had not done enough to secure their consent. "How could you allow yourself to accept this honor on Independence Day, which we are so sensitive about? Are you insensitive?"

One person called for me to be dismissed, but retracted this position after hearing the facts. Another consideration for them is what the media will say about them. This question of sensitivity is very interesting—the ability to give ground, to accept the other's needs. If you feel that you are strong, giving ground doesn't hurt you. The Arabs may come from a less strong position. They may not yet have reached the point where they can give ground, although some of them may be at that stage. Most of them feel that if they accept our need it is as if they are "giving in to the Jews." It is hard for them to accept a Jew as a victim. Some of them cannot sense yet that it is I who need support.

Diana

We are two peoples who have been in a long conflict for a hundred years. The Jewish side, which is the strong side, allows itself to relate to its own history, collective historical memory, roots and

> belonging, and yet at the same time denies our right as Palestinians to do the same.
>
> The Jews feel threatened when we talk to them about issues that are sensitive to them. They feel threatened when we talk about collective memory, or refugees, or the fact that we are a people with roots in this country. The strong, dominant side makes us feel as if we were superfluous here in our homeland.

Since both sides come to the encounter from a society that is living through an unequal conflict, one of the most important assets each side holds is its sense of justice and victimhood. In Neve Shalom, as in other frameworks that permit a discourse of victimhood, the Arabs are in the stronger position. Both sides believe that the reality in Israel is that Arabs have been and still are discriminated against, and that their rights, culture, property, and lives are violated.

When Jewish and Arab groups meet at the Peace School,[14] the encounter often ends in catharsis. The Jewish side, with its heightened awareness of its part in the suffering caused to the Arabs, reaches the conclusion that there is a need for greater equality, justice, and so on. One of the frustrations felt by the Arab side is that after the workshop ends, the Arabs return to the same reality of discrimination. There is no connection between the reality that is created in the workshop and the everyday reality of life.

Some of the Arabs in the village harbor fears that despite the egalitarian experiment, external reality will lead to the emergence of inequality in the community and the school. As a result, attention is given to every detail. Every festival that is a regular occurrence in another school entails a chain of examinations and thorough preparations so that no side will feel offended. Despite the discussions and open conversations among the teachers, there is always the risk that someone will be offended, particularly when the events relate to symbols that are associated with the national myth and ethos.

> Diary
>
> Event
>
> A pre-school teacher distributed a picture of children from all three faiths. The background bore the legend "Jerusalem." A teacher pointed out to her that the picture was making a political statement in favor of the unification of Jerusalem—an issue that is still the subject of negotiations between the Palestinians and Israel.

> Event
>
> A pre-school teacher in the young class handed out Israeli flags to the children (just Israeli flags). Some of the Arab parents were very angry about this.
>
> Event
>
> Children from East Jerusalem told their teacher that they throw stones. The teacher told them that this is dangerous. He told them that when he does reserve duty and gets into this kind of situation (stones are thrown at him), he feels a great sense of danger. Several teachers pointed out to him that the children might potentially interpret this as meaning that he may be the soldier that will fire at them.
>
> Event
>
> An Arab teacher is addressing feedback from her colleagues after she used the word "massacre" to refer to the events of September 2000, during the first days of the Al-Aqsa Intifada, when thirteen Arab citizens of Israel were shot dead.

These representative events emphasize the difficulties and challenges facing the teachers in Neve Shalom. Language and content must always be directed to both populations. Language or behavior that would be considered routine in mononational frameworks (such as distributing flags, discussing experiences in reserve duty, or accusing the aggressive side of a massacre) become highly charged in the binational context. Misunderstandings are inevitable in a framework that brings together populations in conflict. Not every teacher is a diplomat, qualified mediator, or anthropologist. The teachers are good professionals who are willing to teach in an environment that comprises significant ambiguity, and to receive constant feedback relating to language, symbols, and the way culture is presented to the binational group. At discussion workshops, the teachers raise emotional points relating to incidents wherein they have been offended during the course of their everyday work rather than have abstract discourse about events. Thus when events such as those described above occur, the school learns from them. The process of clarification creates a proper behavioral infrastructure for coping with similar cases in the future. Every teacher in the school is required to undergo a process of change, and as time passes, the teacher acquires more tools for coping with the ambiguity that characterizes life in the school. Diana emphasizes, however, that "there is always a feeling that whatever we do will never be enough, and we have to give more time and acquire more tools."

The Child in the Center and as Part of the Group

Some of the residents of Neve Shalom—mainly Arabs, but also some Jews—would like to see the process of study in the elementary school as a one similar to the type of process that takes place at the Peace School. The approach at the Peace School argues that each spokesperson in the group is a representative, a voice of his or her national group. The individual exists only as part of a national group. The emphasis is on the work that needs to be done to develop the insight of the group.[15] Sometimes the Jewish side experiences a sense of identification and catharsis when it understands its role in creating the victimhood of the Arab side. The Jews do not always reach this conclusion, however, and in such cases the Arabs feel frustration at the Jews' inability to see their role in the aggressor/victim equation.

Diana, Boaz, and other teachers feel that the Peace School model is less relevant for the elementary school, where the staff and children are expected to live alongside each other for an extended period. They hold this position despite the fact that the approach of the model has been proved valuable in work done with numerous groups that have participated in workshops.

Diary—A Meeting with Diana

Diana discusses her personal experiences at the Peace School. She reflects on the difference between the current style of counseling in this institution and the development of the desired identity of the elementary school.

I am very familiar with both institutions. I worked at the Peace School from 1984 through 1990, and I even managed the school for two years. At the end there was a kind of crisis, and I left the Peace School along with a few other members of staff. We had an educational and ideological approach that we tried to defend. We undertook action research and developed an approach that enables people to live with the conflict. The researcher was our consultant. Part of the time we had to work on ourselves. We felt that if someone doesn't experience something themselves, it will be difficult for them to pass it on to others.

I can still see this point with teachers today. People who open up and ask themselves questions, even questions that are supposedly forbidden, can undergo an internal experience and develop. But during this process of exposure, the teacher is unprotected. They need extensive support, and through our work we supported the teachers greatly.

The perspective of the Peace School is problematic, because above all I am my own representative. The same is true of the student, who is his/her own representative, as well as the representative of his/her culture. The student should be free to present their own personal position. I am not a representative of my people all the time, because that approach negates human individuality and blocks possibilities. According to the current approach of the Peace School, I cannot be myself, only part of a group. But the truth is that I am a bit of both.

I was born into my national identity and it has become part of me. I have learned to be proud of being a Palestinian, while at the same time examining certain foci of my identity in a critical way. Every individual has many identities, some of which he or she chooses while others are there from birth. We have to live with this diversity. The fact that there is no Palestinian state enables me to emphasize my Palestinian identity above my other identities. The situation should not deny anyone their right to live with their multiple identities, just as I do.

Diary—A Meeting with Boaz

As in many other schools, we also put the child in the center. But in Neve Shalom there is a certain tension, because it is obvious that the child forms part of a group. The place of the group within the school is very important. There is some tension regarding the question as to what is more important—the group or the child. Reality shows that while the child is indeed in the center, as opposed to schools where that is the main motto, here we have to pay more attention to the collective and national issues. Every child comes here with their own reality, and the reality is that the Jews are dominant. How can we ensure that this reality will not be perpetuated here as a matter of course? The Peace School runs workshops for older children and adults, and they have chosen to work in groups. There, the Arab side has to beat the Jewish side in a verbal battle and prove who is the victim and who is more moral and just. The Jews will try to narrow the gap, but in this struggle it is obvious that

> the Arabs are the victim. As the Arabs grow up, they gain a sense of strength, overcome their feelings of inferiority and free the Jews of their sense of superiority and occupation, at least in the counselors' imagination. While the workshop takes place, the participants' feelings change. Afterwards, the Arabs return to their daily reality and have a sense of disillusion and disappointment.
>
> This process is impossible here at the school. We do not use this kind of tools here. I can't say here that Ahmad is the Arabs, that isn't appropriate at this school. So the tools they use at the Peace School are not right for us. We don't run workshops here. We are not going to base our studies and education on a constant confrontation between groups. That isn't suitable for a school that educates to tolerance, peace and pluralism. But some of the Arabs in the village want to use these tools come hell or high water. There may also be some Jews who feel that way. In various discussions in the classes, such situations sometimes emerge, but in the final analysis, the purpose of the school is to allow each child to grow, to develop their own holistic identity and to provide them with tools for coping with life.
>
> Our goal is to give each child a confident identity so that the other side seems less threatening. It is precisely the otherness of the other side that enriches the student, as well as the fact that each student and their culture are valued and given as much space as possible. We try to provide less harsh answers for people who are looking for a harsh answer. Equality does not mean that one group submits to the other. A child who has developed a confident identity, learned to live with the other and to cope with reality from a sense of equality will internalize this identity in their way of life and will not submit to reality.

In the ideological debates in the school, we find echoes of the critical pedagogies of Freire, as well as ideas of Dewey on humanistic educational approaches and postmodern approaches. Both the Peace School and the elementary school in Neve Shalom offer a revolutionary educational approach by bringing together two distinct populations to learn alongside one another. The dialogue that is created among the staff, and the dialogue between parents and children, is critical in nature. The teachers are in agreement that the status quo in Israeli society is not right for us today and must be changed. Freire argues that dissatisfaction with the status quo and the desire for change are good catalysts for meaningful educational activities (Freire, 1994). The willingness to be exposed, to engage in discourse on the dominant power mechanisms, and to create parallel and alternative narratives recalls the postmodernist tendency to attack and dismantle metanarratives. However, the school offers an alternative ideological blueprint based essentially on humanistic and democratic ideologies. Postmodernism provides a tool for dismantling power mechanism, but not for the total dismantling of any ideology. The school is not left in an ideological vacuum.

The school lives with the tension between more neutral approaches, such as those of Maslow and Rogers, which emphasize the centrality of the child, the importance of his/her emotional development, nurturing self-image, and self-acceptance, and theories that advocate a more political and militant political discourse in the spirit of Freire and Giroux. The compromise offered by the elementary school in Neve Shalom is based on two main axes: firstly, the conflict as a method for effective study is not appropriate for such young children. This style of learning may be better suited to adolescents and to adults who have already developed their identity. The elementary school offers a preparatory stage for such potential encounters, enabling a positive multicultural encounter with an emphasis on national and cultural identity and on accepting the other. This approach is based not only on consideration for the nationality of those involved, but also on consideration for their developmental stage. As and when a junior high school and a high school are established in Neve Shalom, I believe the study model may move closer to that of the Peace School in the village.

A second axis, no less important in mitigating militant and critical tendencies, is the community of parents. There is a limit—one that is, perhaps, particularly evident among Jewish parents, specially those who do not live in the village—to their tolerance to engage in sometimes overly harsh dialogues relating to the symbols and ethos of Jewish society. It must not be forgotten that for many of the Jewish parents, their choice of Neve Shalom is a choice of a good school, along with their identification with the pluralistic and democratic ideology behind the encounter. Teachers at the school fear that if the education is perceived as too radical, parents may withdraw their children.

Interview with MB, a Jewish Mother

We live in Zur Hadassah (a suburban town near Jerusalem). We used to live in Jerusalem. This is the third year that our children are learning in Neve Shalom School. We didn't want our children to study in Zur Hadassah. We heard about Neve Shalom, and our choice was between this school and a school that only just opened in Zur Hadassah, where the students walk to school. The journey time from Zur Hadassah to Neve Shalom is about half an hour, which is significant. But if you want a high-quality education outside Zur Hadassah, then anyway our daughter would have had to travel to school. I am sure that there are other parents who also chose Neve Shalom as an alternative to a mediocre local school. Today the school in Zur Hadassah is developing and improving, and the advantage that it is close to home has become more significant.

When we lived in Jerusalem, the children went to a regular school. There were forty students in each class, with all the limitations that brings. Apart from that, the ideology of the school in Neve Shalom matches our approach. The conditions—small classes and innovative studies—are completely different. Students don't get anything out of a nature class consisting of a walk around Pisgat Ze'ev, the neighborhood of Jerusalem we lived in. There's nothing to look at, not even a leaf. We also had practical reasons for moving out of Jerusalem.

The children here meet other children. They hear about their festivals and get to know other cultures, the other people that lives here in this country. They learn that "Arabs" is not just a concept, but real people. That gives them values. These are things that were only theoretical for us. The Arab and Jewish children go to play with each other in the afternoons, although the distances are a problem. My daughter meets more with the Jewish girls, because although there are joint activities, the distance plays a role. The same thing applies to Jewish girls who live a long way away. Every few months they meet on Saturday or during the vacations, but usually they hang out with the children who live closer to us in geographical terms.

They related to the Intifada from the first day here. They spoke to the children, and sent reading materials and letters to the parents. The children learned to talk about their feelings. Today the situation is beginning to settle into a routine—maybe that's a bad thing. It was difficult at first when the Israeli Arabs were involved in the events. We were rather concerned at the lack of balance in the explanations. For example, it was claimed that our soldiers were shooting for no reason. But on the whole, the children find it easier to cope with these problems.

As far as I know, no Jewish parents have taken their children out of the school recently. The Intifada has actually made people close up within the school. We had some problems with the 6th grade class this year, but this is a class that has had some social problems for a few years. There were almost no children from Neve Shalom in the class, so many of the Jewish children left the school.

Sometimes a child decides to leave. Sometimes the children have friends from their home environment, and it is easier for them to meet. There is also a social dimension—it is easier to make friends with children from the same area.

Recently, a national summer camp was held for Jewish and Arab children. A large number of children came from all over the country, including some participants from Neve Shalom. We saw that the place makes a difference. There is an incredible sense of unity among the children from Neve Shalom. They have become a unified and consolidated group of Arabs and Jews. All the other groups arrived as mononational groups, and the great integration in the group from Neve Shalom really stood out.

The school is perceived as a good one in academic terms. After finishing the 6th grade, the children can get into whatever school they want.

I assume that parents who send their children to this school are left-wing, but in different shades. You can meet all kinds here. A lot of the parents have bachelor degrees and above (she and her husband both work in prestigious professions). If you decide to move your children to a school like this you have to have some kind of awareness.

The Arab parents definitely have a high level of political and cultural awareness. It isn't a neighborhood school that everyone in the area goes to, although there are also parents who are not well-off who send their children here.

The parents do not have many opportunities to meet other Jewish parents or the other national group, just because of the daily race that we're all involved in.

This representative interview illustrates the tendencies and motives of many Jewish (and Arab) parents who send their children to the school in Neve Shalom. The school constitutes one of many options—Tali Bayit Vagan, Keshet School (which is described below), the Experimental School, and so on. For the staff, principals, and the residents of Neve Shalom, there is no immediate alternative. The decision to live within the conflict is final, for better and for worse. The Jewish parents from the immediate vicinity and those from distant places choose the school because it is considered a good, intimate, and humanistic school, even if they have some concern about the school's excessively radical statements based on pedagogy of liberation and struggle. The school cannot exist physically without these children, and when it discusses complex and controversial issues, it is obliged to take into account the potential reactions of their parents.

Accordingly, two factors—the age of the children and the position of the parents—moderate the critical message of the school and strengthen its humanistic, democratic, and universalistic message. The school develops a sense of empathy and understanding among the children. They perceive the other side as humans and to a surprising extent adjust their abstract positions regarding the other. Sometimes they even influence their parents.

Language and Equality

One of the most painful issues for the Arab members of Neve Shalom in general, and the school in particular, is the question of using the same language. There is no problem making decisions to advance equality in the village or in the school, and it is possible to appoint more Arabs than Jews to key positions, but it is hard to motivate the Jews to learn or speak two languages. The effort needed to achieve this transformation seems to be insurmountable for the participants from the various communities in Neve Shalom. Although the school was declared bilingual from its inception, there can be no doubt that Hebrew is the more functional and dominant language. In this respect, the school reflects the broader reality of Israeli society.

Hebrew is the majority language in Israel, and it guarantees advancement, education, and integration in society and the workplace. A high level of Hebrew, combined with good English and an ability to use computers, is a basic tool for economic success. Arabic is not currently seen as a significant tool for socioeconomic success (although in the future, in a hypothetical scenario of peace leading to the opening of regional markets, this situation might change). The English language also reflects the cultural orientation of Israelis toward Western culture, which is perceived as closer and more relevant than Arab culture.

By contrast, the Arab side attaches great importance to learning Hebrew and English—languages that Arabs, just like Jews, need if they are to prepare themselves for work and life in a society that mandates an open economy and ties with the West. Arabs gain power in Hebrew-speaking environments, such as the universities. They write their doctorate theses in Hebrew or English gain strength, and change through their contact with the West. They do not usually have a functional need for Arabic as a tool for success.

Many Arabs in Neve Shalom are worried at the declining standard of Arab and deplore the introduction of Hebrew slang into their language. They see their language as a cultural asset that is under threat. Their predicament is reminiscent of the situation of California's Spanish-speaking population that advocates bilingual education and emphasizes the importance of its language as a tool for preserving its identity, heritage, and culture. However, there are many who oppose this trend, including members of the Latino community itself. The famous teacher Escalante,[16] who managed to enable a large group of Latino youths from an underprivileged neighborhood in California to a very high standard in mathematics, has spoken out against bilingual education. He argues that this prevents the youngsters from acquiring knowledge in English, as a result of which they are unable to integrate into society. He claims that it is better to pay the price of losing one's mother tongue in return for the potential strength of gaining from a thorough knowledge of English and the sciences.[17]

Diana sees the question of language as one of the most important and difficult challenges facing the school.

Diana

If the Israeli authorities argue that Arabic is important enough, because we are part of the Middle East and must take account of those who have become a minority in this country and who speak Arabic, Arabic will become a legitimate subject in all schools and there will be a change.

We are talking about learning Arabic together with its whole cultural context, since the school cannot create change on its own. The children are part of the external reality, which does not provide sufficient encouragement to learn Arabic. The Arab parents are also very concerned to provide their children with the tools that will help them survive in a competitive society. It's a problem.

Studies are currently underway to find ways to help Arabs to love Arabic and to continue to speak the language. The Arabs in Israel are slowly internalizing the fact that it is only through the use of Hebrew that they can be successful. I cannot survive at university using Arabic—even Arabic is taught through Hebrew. It's absurd, but that's the way things are. It's a pity.

There are many models around the world for bilingual education, for example with Spanish- and French-speakers. All the models emphasize the need to devote 70 percent, or even more, to the minority culture and language.

Because of the complex nature of a Jewish-Arab school, we have not so far allowed ourselves to internalize perceptions from the Western world. We still cannot devote 90 percent to Arabic in the early grades, or even 70 percent as is the case in some schools abroad that begin by emphasizing the minority language. We teach half-and-half here, and even that is difficult, because the Arab teacher translates what he says into Hebrew. The students hear more Hebrew than Arabic. This point must change quickly.

If the Jewish teachers could speak both languages, we could begin a process of change. That's part of the problem. Another part of the problem is the attitudes of the Arab teachers. Are they really convinced that the Arabic language deserves respect and can be used alongside Hebrew? I think this is a process that we should all undergo. If I am confident in my own identity and proud of my affiliation, I will not hesitate to nurture my mother tongue.

A Discussion in the Staffroom

Arab Teacher

One of the problems is equality in language. The Jew who knows Arabic best knows less than the Arab who knows Hebrew worst. (He refers to teachers)

Jewish Teacher

Some of the Arab teachers are angry because staff discussions and other events take place in Hebrew, and this becomes a tool for getting at people. The subject of language can become an endless source of conflict.

Boaz

There is no equality when it comes to language. Until now, we have worked toward the idea of a 50:50 equality. But the experience of bilingual schools in the United States (Spanish and English) shows that you need 30 percent versus 70 percent, and sometimes even a wider gap. We still cannot suggest such a division, because for the Jews who come here from Neve Shalom and elsewhere, even 50 percent is seen as a compromise, and the parents call our school any time something bothers them. The parents are going through a process now that will prevent us losing our "customers." In the beginning, when the school was smaller, we could possibly have started speaking Arabic, but it is harder to introduce such a process now. This is why it is hard for us to find equal numbers of Jewish and Arab teachers. Maybe in the future we will adopt a funnel model, beginning with 90 percent use of Arabic at the beginning and slowly reaching 50 percent at an older age. But this would change the work of preparing the staff and the balance between the languages. There is a problem here, but we can cope with it. It means dismissing teachers. My own experience suggests that even if there are more Arab teachers, the Jews do not necessarily improve their Arabic. We need to develop a program that will meet this need.

Diana

The atmosphere has been more difficult recently because of the Intifada, but the truth is that it has been here all the time. We were born into the conflict. It's easy to say, "It's not me, it's them." Language is an important factor. The Arab parents and teachers and myself know both languages, but the situation is different on the Jewish side, so you already have a gap. We need to change the situation, bridge the gap and increase the level of equality. We haven't been successful in this. Jewish teachers have learnt Arabic and use it a little in class. It is important for the teachers to work in Arabic, because they provide a role model for the children.

We had a problem with the Arabic textbooks. They were so boring that I almost threw them away. But the teachers still feel that they need them. Students knew how to read and write in Arabic, but could not speak the language. Recently we changed the system: the students wrote in Hebrew and spoke Arabic, and then they became more interested in learning Arabic. This method is controversial, but over the past two years, the Jewish children suddenly like Arabic. The next question is how to help the Arab teachers speak less Hebrew and not to translate so much of what they say. We are trying to work out how to make equality work in the field of language.

In the case of the previous issue, which related to the tension between collective education and the emphasis on the individual, the pedagogic direction was determined by the parents and the age of the children. On the question of language, however, the main consideration is the parents' position. The nature of the demands depends on the perspective and observation of the one who is demanding. As someone who has regrettably taken no time to learn Arabic, I was very impressed to see the children playing together, the Arabs speaking Arabic while the Jewish children answered in Hebrew or in a mixture of the two languages. Arabic is a functional language in Neve Shalom, albeit not as functional as might seem desirable to the Arab observer.

> Diary
>
> The school day ends again. The children shout in Hebrew and Arabic. Soon the school will be empty. Some members of staff hold a discussion by the secretary's office, and walk toward the staffroom. The conversation takes place in Arabic. A Jewish child waits by the secretary's office. It is unclear whether he will be going on the school bus. He calls his parents from Diana and Boaz's office, tells them that he fell down, but it "doesn't look broken." His mother tells him to come home. The secretary talks to him in Arabic. He seems to understand everything and answers in Hebrew. I go to the staffroom. The air-conditioner cannot compensate for the hot day. The teachers look at me for a moment and continue their conversation.

To summarize the question of language, it seems to me that in contrast to the usual situation in Israeli society, where Arabic is an absent and unrepresented language, Neve Shalom aspires to equality and to the proper and equal representation of both languages. However, since the learning process takes place not only inside school, but also outside of it, education is through the totality of experiences a person undergoes (Harrington, 1979), the school cannot be expected to completely change the status of languages, although it certainly manages to improve the situation and give the Arabic language a dignified position. Cautious changes in the curriculum, made using the lessons learn from the experiences of bilingual schools around the world, might improve the knowledge of Arabic among the Jewish students. Political changes in the region may also make Arabic more attractive, if only from an economic and utilitarian perspective.[18]

The Teachers in Neve Shalom

From the interviews with Boaz and Diana, it emerged that the ideal teacher at the school is someone with a high level of self-awareness, one who is politically aware, creative, and has a high level of emotional intelligence and an ability to act in situations of uncertainty. However, no education system works with an ideal staff. There are always factors that delay the realization of the ideal. Just as equality between two languages in Neve Shalom is an aspiration that is realized only partially, success in getting teachers who understand the idea has been limited.

> Diana
>
> From a certain point, the school began to grow. We hired teachers who tackled with the real problems, but we also hired teachers whom we wouldn't have accepted if we had the choice. It took a long time for them to settle in, if at all. They may have come to us because they couldn't find a job elsewhere. Some of them tried hard to work on themselves and change after they were accepted here.
>
> We need very creative and open-minded people who are willing to change and to be changed. About half the teachers are like that, but it is hard to teach here if you aren't politically aware, because we relate to that dimension. The teachers have to know what to keep to themselves and what to convey to the children, because events happen every day. We teach current events and how to cope with the current situation. We teach exercises that will help the children express their feelings, develop the ability to listen and live in a complex reality.
>
> We let the teachers down by failing to provide proper training in how to cope with the complexity of work in the school, because there is nowhere else that can train teachers to meet our goals. They need a significant experience at least once a week. It's true that we have a staff meeting on Sunday for an hour and a half, and that's very important, but there are always lots of routine problems and issues that need to be discussed. There isn't enough time for reflection and proper training. The teachers are very sensitive, though, and they do their best to meet the complex demands of the school. There is something that binds the teachers to the school. Their work here influences their soul and improves their self-esteem: "I'm a teacher in Neve Shalom." Their work experience here will help them look at any problem in life from a different angle and not to take anything for granted, so they can see the different shades in educational work. The teachers absorb this message, but they sometimes lack the tools to realize it.

For a year, I observed the teachers work in the school, in workshops and classes, and interviewed several of them. I believe that the majority works to achieve a high standard out of a sense of mission. Most of the teachers experience a process of change in themselves, although it is not always possible to identify the particular stage that each teacher is at. The work in Neve Shalom School is more complex than in a regular school, because of the sensitivities that stem from cultural differences. It may be difficult for the principals of the school to accept that some of the teachers are at a much earlier stage of exposure to political issue and coexistence. However, a real effort is made to enable these teachers to reach a more complex stage in the process of change, through workshops, encouraging discussion, joint reflection, and similar opportunities.

I am convinced that there are teachers who could not teach anywhere other than in school such as Neve Shalom. These teachers are inquisitive and strongly ideological, and they believe that working in a school with a sense of mission and purpose, such as this school, compensates for the low status and poor salaries of the teaching profession.

Conclusion

Diana

Despite the difficulties here with the issues of excellence and achievement, I meet a growing number of Arab parents who state that the school emphasizes the nurturing of their child's personality, and that this is very important for them. One mother told me: "I know that the school doesn't focus enough on achievement, but I am more concerned that my children feel confident and know how to express themselves. You develop their personality so they can gain strength in life."

If the child feels confident in their identity and heritage; if we have managed to educate them to accept the other and those who are different and to overcome barriers; if we have managed to give the child an educational experience that is consonant with their learning style—then we have succeeded in our task here at Neve Shalom / Wahat Al-Salam.

Diary: The Students' Visit

We sit in a comfortable room with a spectacular view. The student teachers observed activities in the class, and now Diana is having a concluding discussion with them. She emphasizes the issues of equality and language. Occasionally one of the students asks a question. These have been an intensive four hours, but the students still seem very interested in Diana's explanations. Many of the students are not used to meeting Arabs who function in positions of power. Most of them would probably not consider Neve Shalom as an option for their own children, but their questions show that they strongly admire the educational and ideological effort invested in the school.

Boaz gave an opening lecture, and they met Diana at the end of the day. I see that Mali (not her real name) is very agitated. She comes up to me later and says: "I was very moved by what Boaz had to say (he spoke about the death of his son Tom). I went up to him and told him how much I felt for him, because I also had a uncle who was killed in the war. I cried, and he also started to cry. We stood there for a few minutes. Then I told him that although I don't accept the position of this school—I am a different person from a different background, Mizrachi, religious, maybe right-wing—I still understand what you are trying to do here."

Boaz

In this era of terror attacks, a child comes here from the regular education system and tells us that his mother orders him to come home as soon as pre-school finishes, or else an Arab will catch him. There is awareness of the conflict and war here in Neve Shalom, but the student here can go to their class teacher and talk about the terror attack. This actually happens. The child asked the [Arab] class teacher, "Are you a terrorist?" Although she was young and could easily have been offended by the question, it is typical of this school that she used it as an educational opportunity (the boy's father is a disabled war veteran). This is great. A child like that is going to test his hypothesis: the terrorists are Palestinians; you are an Arab, maybe a Palestinian; so are you a terrorist? It sounds logical. You can hold a discussion in class, and the child could be actively suspicious and search her bags. But even so he feels that there are Palestinians whom he can trust one hundred percent. This is something concrete in his life—it isn't just a vague statement that "They aren't all like that."

Neve Shalom is not a natural part of Israeli society. The natural situation in Israeli society is one of inequality, racism, and oppression. It is remarkable that there are people who are willing to establish a community that is so sharply different to the dominant Israeli ethos, particularly since the general atmosphere and environment present numerous difficulties. There are many organizations in Israel that encourage coexistence and encounters, but there is no parallel to this school community whose members have chosen to live and fight the conflict together and not just to talk about it.[19]

Neve Shalom School offers a framework within which the participants can engage in constant negotiation with their values. Since there is no role model for such an experiment, the participants create the model by themselves through trial and error, self-examination, learning from experience, and acting within a world of change. There is an expectation that eventually the staff will be successful, will be proactive, and will help create an alternative curriculum for the school. Such a curriculum draws on the cumulative experience of life in the village and at the Peace School, and on the personal narratives of the participants, particularly the staff and the children.

This curriculum is innovative and revolutionary. The Ministry of Education does not have programs suggesting ways to cope with the simultaneity of Independence Day celebrations and the commemoration of the Naqba. Neither does the ministry have any suggestions as to how to mark three festivals of different religious within the same school, or how to run a bilingual school.

The school in Neve Shalom is not merely a place where children are taught. As we have seen, it is also a place where adults learn from children. The surprising observations of the children underscore the educational potential of binational schools. The children evidently undergo a process of socialization around "soft" myths, in place of the dominant "hard" myths of Jewish and Arab society—myths that have grown more potent as the Intifada has continued. The "soft" myth offers the participant pride in his/her own national and cultural identity, while refraining from negating other myths. Moreover, it invites and encourages the participant to be a guest in the other's myth, understanding that this does not threaten his/her own myth. It is true that this approach is not without its problems, most of which are due to the "baggage" that adults bring from the past. The adult participants were not shaped at a place such as Neve Shalom. Past grievances sometimes make it difficult to soften the myths, and on occasion it is the children who help the adults to appreciate the destructive results of aggressive and stereotypical myths. In the future, it will be interesting to examine on all levels the personality and identity of the graduates of the Neve Shalom experiment.

As mentioned above, in the future the school hopes to expand its mandate, establishing a junior-high and even high school. The school also provides a consultative framework for similar programs implemented at various locations around Israel, although the model is not replicated in a complete and identical form. The transition from the status of an experiment to the official status of the Ministry of Education and the efforts to found a junior-high and a high school are exhausting processes that require extensive efforts by the staff. It is to be hoped that the Ministry of Education will see this school as a model to learn from, and not just another school to be supervised.

The Peace School in Neve Shalom and the elementary school are making an important contribution to an increasing understanding of the Jewish-Arab conflict—an understanding that creates the possibility of coping with and changing the conflicts. There is a difference between the two frameworks, however. Whereas the Peace School uses Western tools taken from the field of group psychology, particularly ones relating to coping with problems of identity, minorities, and racism, the elementary school operates according to a pragmatic philosophy reminiscent of the vision of "Mediterranean enlightenment"—the term used by Ohana to refer to thinkers such as Camus, Memi, Ben Jelloun, and Mahfouz. Ohana defines Mediterranean enlightenment as an idea that seeks to achieve the "universalization of the human race, which, despite its myriad faces, forms the family of humanity. In the Mediterranean context, this ideal goes beyond abstract enlightenment defined by intellect, ideological sermonizing and philosophical sophistication. Those who embody this ideal have internalized warm and immediate contact through dialogue between cultures, through connecting to the different, the foreign and the other."[20]

In other words, the elementary school, through its ongoing work with the children, is involved not in abstract philosophizing and theorizing on the subject of encounter, but in developing a practical and simple philosophy that aims to meet the naive but pertinent questions raised by the children. The type of dialogue created at the school between adults and children is similar to the straight-talking dialogue between Ben Jelloun and his daughter in *Racism Explained to My Daughter* (1998).[21]

Of all the rifts that threaten Israeli society, the national rift between Jews and Arabs would seem to be the harshest and most dangerous. It is encouraging to see that, despite the difficulties, there are those who are willing to engage in serious educational work at the grassroots level, even before the seeds of suffering and hatred have had a chance to develop. The school has undoubtedly attracted international attention due to its courage and its innovative message. The Ministry of Education has embraced the school and begun the process of granting official recognition. The ministry would do well to regard the school as a model and a base for learning. Educational frameworks can exert influence. Student teachers should learn about the origins of the Jewish-Arab dispute, and Jewish students should learn more about the Arab minority that lives in Israel. As teachers of the future, it is important that they visit Neve Shalom and deal with the questions raised by the educational activities of the school. This does not mean that every student will change his/her political opinions after visiting Neve Shalom, but as agents for socialization, teachers should be exposed to different models of schools as they move into the educational field. Such models, whether they arouse questions, agreement, or reservations, will help the students develop a more profound and complex teaching style—something that can only benefit their students, and—in the long term—all of us.

CHAPTER TWO

Bialik School

(Grades 1st–6th, approximately 270 students)

Diary

I travel to Tel Aviv in a comfortable and air-conditioned bus. The school is a two-minute walk from the bus station. As I leave the station, I see people walking purposefully, while others seem to be wandering with no clear goal. Some people are sleeping on benches. The shop windows bear signs in Russian and other languages. Nearby, pubs offer a few moments of relaxation to workers who have finished a hard day's work, as do the adjacent brothels, which keep women in conditions of slavery that is ignored by Israeli law.

I walk through the streets, listening to the jumble of languages and watching people of every color and nationality—a foreign city within the first Hebrew metropolis.

Native Israelis who cannot afford to move out have stayed in the vicinity of the central bus station. At one stage, they were convinced by the argument that the massive white elephant of the new bus station would revive the area, leading to a rise in property values.

Disappointment and neglect seem to ooze from the buildings, which are coated in thick grime.

The native Israelis who live in the area have been joined by other residents: New Israelis who have arrived in the country recently and face economic hardship, as well as tens of thousands of migrant laborers and some Palestinian collaborators. The collaborators are the victims of the same dispute that has prevented most Palestinians from working in Israel, creating the incentive to bring migrant laborers from across the world as an alternative source of cheap labor. This process has created a social problem in Israel that has exacerbated as the years have passed.

I walk toward a school that provides a pinpoint of warmth in this sea of alienation and attempts to bridge the gulfs between these disparate population groups. The guard at the gate does not recognize me yet, but hesitantly permits me to enter.

I arrive during recess. Children from around the world are playing together. I cannot see any evidence of violence or bullying. I notice that the children play in mixed age groups, with a relatively high level of interaction between boys and girls. The sounds of the children's playing sometimes manage to drown out the noise from the busy street.

Amira Yahalom, the principal of the school, is waiting for me in her office.[1] She is elegantly dressed and smiles often.

Amira is responsible for making the school what it is today. The school has changed along with the neighborhood. No official decision was made to transform Bialik into a school for the children of migrant laborers, but the parents arrived at the school, registered their children, and gradually they came to constitute a majority of the student population. When Amira took over as principal, the Ministry of Education considered replacing all or some of the teachers to reflect the changes that were underway. However, replacing teachers was not consonant with Amira's educational philosophy. She argued that "educators should be given a chance to prove themselves, with the assistance of a professional and supportive process." For some of the teachers, the process of change was personal and painful. Not a single teacher was dismissed, reflecting the belief that every educator can learn to teach and work differently. The result was the development of a perception in the school that change must come from within.

Warmth and Acceptance

Amira and the staff she heads believe that teachers must work intensively in the emotional realm. In every class, strict attention is paid to ensuring that the teachers provide a homelike atmosphere for children who are almost homeless. The school also meets those needs that are in the case of more prosperous populations met outside the school. The parents of the students at Bialik School work hard, sometimes from morning to night. There is a large proportion of single-parent families. Deprivation is widespread, and the teachers often find themselves playing the role of parents. The level of emotional acceptance demanded of them is high, but the children respond by showering love and attention on their teachers, an exchange that I have rarely seen in any other school.

Diary

During recess, I walk around the school and observe the activities in the classrooms. Some children prefer to remain in class during recess. Some of them draw on the blackboard. In the 5th grade class, a Romanian girl draws complex geometric patters on the board. S, a Palestinian girl from Jenin, writes a letter of thanks and love to her teacher: "To G, thank you for what you teach me."

Both girls tell me that they love their teacher G, as well as S, who taught them the previous year (although one of them adds that S used to shout a lot).

A closet in the staffroom is full of used clothes that the teachers have brought from home. Two girls come into the staffroom, look through the clothes and choose items from neatly-stacked piles. Their behavior in the staffroom is reminiscent of the behavior of children in their own home. When they leave, a teacher remembers that she has a curtain she wants to donate. Several organizations also donate clothes to the school.

D, a teacher in the 6th grade, walks along the corridor. I follow her. She meets P, a young girl from the 1st grade. P is very agitated and angry, and D attempts to find out why. It emerges that a dispute occurred among children in the neighborhood. D takes P to her teacher to discuss what has happened. P calms down somewhat as she talks to her teacher. She mentions that her mother is in hospital, and her teacher is concerned: who is looking after P? When I meet Amira later, I mention how surprised I was by the level of the teachers' involvement in their students' problems, even when these are outside their field of responsibility.

Amira

The teachers here know each student personally, but it is impossible to be on top of everything that happens. We did not know that the girl's mother was in hospital, so we could not understand why she was in such a bad state. The family's social worker came to speak to us, and then we realized what was happening. The family

> is going through a very difficult crisis. It is good that P stays at school until 3 o'clock. In any case we are contacting the relevant authorities and together we will find appropriate solutions.

The school appears to function according to Maslow's scale of needs.[2] Maslow explains that it is impossible to achieve cognitive and spiritual growth, self-respect and self-realization without addressing basic needs: nutrition, protection, shelter, and emotional response. The school seeks to meet these needs, since the difficult and complex social and economic conditions of the children's parents prevents them from taking on this role. When the parents' condition is problematic and the school fails to meet the children's physical and emotional needs, the damage to the children is irreversible, as described in graphic and powerful terms by Kozol[3] in "Savage Inequalities: Children in America's Schools." Kozol argues that an unequal education system that is unable to provide children with equal opportunity (and perhaps does not even seek to do so) perpetuates social gaps.

This is not the case in Bialik School. Given the demographic profile and the prevailing atmosphere of malaise and despair, the school might be expected to be headed for failure. Instead of failure, however, the school has nurtured faith in its ability to achieve meaningful change in the lives of these children.

> Diary
>
> I sit with N, a teacher at the school. During a class for a group of 3rd grade children, she tells me:
>
> I work with each student according to their individual intelligence and background. With the current 3rd grade group, I am still working to instill basic reading skills. They face cultural and environmental gaps, but we give all of them love.
>
> One of the children makes a comment that I cannot hear. N asks him, "Do you want to come up and teach instead of me?" He agrees, and she encourages him. He comes up to the blackboard, writes the first three letters of the Hebrew alphabet, but has a problem with the next letter. Some of the children try to help him, while others take pleasure in his mistakes. The teacher smiles. She begins to teach them another rule of Hebrew orthography; the children join in willingly and clearly enjoy the class. Occasionally one of the children curses, but the others ignore it and continue with their work. The teacher asks R if he wants to wash his mouth

out. He falls silent and withdraws into himself. Only he and I seem to have heard the remark. N tells the children to open their exercise books and write. They do so, R included. She continues the lesson.

Teacher G

We plan our work as a team so that the teachers are coordinated. I make sure that the material is varied, and try to find special means that are not used in the other classes. I give them enrichment and reinforce what they have learned over the past week. As time passes, each one makes progress in their own individual way. We use a wide range of methods and approaches for language acquisition.

Change and Flexibility

Individuals and human societies generally function well in conditions of stability. The only stable factor at Bialik School is the willingness of the staff to deal with a world that is uncertain, different, and constantly changing. As in Neve Shalom School, the curriculum in Bialik must be adapted to meet the needs of a population that is different from that for which the standard curriculum was prepared. No one in Israel has prepared a curriculum for a multicultural school. Many of the children at the school have experienced the crisis of migration, by themselves or through their parents. Migration creates culture shock and an acute sense of uncertainty. Moreover, many of the students at the school, and their parents, lack the safety net of protection and security that is provided for Jewish immigrants to Israel. Many of the parents have been dismissed by their original employer and therefore face the threat of deportation. As a result, they approach the establishment with caution and suspicion.

The Israeli government is acting forcefully to "reduce the problem of the migrant laborers." One of the ways in which it does this is by locating and deporting laborers. At the same time, Israel is a signatory to conventions on children's rights that require a total separation between the status of the child and the status of the parents and demand that the children's basic needs such as protection, education, security, and the like are met. Faithful to its own ideological approach, the school operates by the same standards, rejecting the belief that it is acceptable to deport families along with their children in the middle of the night, without prior warning.

Amira and the teachers at the school must be patient as they initiate processes of change. They must be aware of the multicultural reality at the school, as Amira notes that

> If you aren't flexible inside yourself, you have to develop this quality. The teachers require massive support from social workers and psychologists. I feel that the formal education system is not always aware that educational staff have these needs.

I asked Amira whether she felt that Bialik School is similar to the other schools that encourage an intercultural encounter. Amira feels that the teachers at schools such as Neve Shalom and Keshet are doing interesting work, albeit in conditions that lead her to call them "laboratories." She agrees that Neve Shalom and Keshet are similar to Bialik School in that they create an intercultural encounter but feels that she is managing to secure her objectives in social conditions that are much harsher. She states that she has readily embraced the concept of the Ideology of Repair, a concept I developed to describe schools that operate in this spirit. In recent years, Amira has been involved in studying Jewish texts in various frameworks for examining Jewish identity from a pluralistic perspective and has found support for the Ideology of Repair (Tikkun) in Jewish texts and traditions.

The school evolves as the surrounding reality changes. Based on their desire to meet the numerous, diverse, and changing needs of the students, Amira and the teachers at the school seem to have developed flexible response mechanisms, extending the definition of the school to a place that is much more than an institution where they study for a few hours a day before returning home to the responsibility of their parents. This school accepts responsibility for various areas that are not part of the usual mandate of Israeli schools. Amira's charismatic personality seems to have helped her develop positive contacts with external institutions that have agreed to help the school in different ways. Her cognitive flexibility also enables others to find creative solutions for various problems faced by the children.

> Amira
>
> Next week, we will meet in the evening to present our activity groups to the parents. There is a community center in the neighborhood. I suggested to the director that he should work to encourage the community in the area to use the facilities at the center. It took time, but now the children are familiar with the center and see that it is not far from their homes and not difficult to reach.

> The parents also sign their children up for activity groups. The community center is subsidized, but the parents still have to pay something. The parents realize that it is important for their children. There is a library at the community center that is open in the evenings. The parents and children didn't know about the library. We tried every possible way to market it, and now they know about it and come there, even though the migrant laborers and the Arabs do not want to be noticed and find it hard to come to places that are associated with the establishment. We managed this by sending a forceful and uncompromising message that we accept you, because you are humans. They can only come here and receive help if there is an atmosphere of mutual trust.

A common criticism of state institutions is that they fail to cooperate or to pool resources. Amira sees a great potential value of creating networks of mutual assistance among the institutions. She brings the children to the community center as part of a project that has become a success story. Together with the staff of the center, she developed a system of subsidized activity groups for all the children. Despite the subsidies, the parents are willing to pay a few dozen shekels for what she offers. More importantly, the parents and children have come to see the community center as a friendly and homelike place, thus helping the it realize its declared mission. Such activities have helped make Bialik a successful community school, in the sense that it pools resources with the surrounding community. This approach sees the school as part of the broader communal system and places the child at the center, acknowledging that the reason why the students are affected both directly and indirectly by the processes occurring around them.[4] We are aware that the reason why disadvantaged populations are sometimes unable to help their children is not a lack of resources in the community, but a lack of knowledge as to how to consume these resources. Bialik School seems to have been successful in bridging this knowledge gap, even in the case of populations that are afraid of exposure to the establishment.

Teachers and the Process of Change

As noted above, the teachers at Bialik did not receive any special training for working in a multicultural school. Most of the teachers were already working in the school before Amira took over as principal. The process of change they have undergone has been complex and difficult, requiring them to adapt to a type of "internal flexibility." Amira's personality has shown them that the task is possible. She did not merely speak about the need for internal flexibility, but she also illustrated it in all her actions in the school. Most of the teachers were bitten by the bug of internal flexibility, which has become a central part of the school ethos.

> A Teacher
>
> We have a normative 2nd grade class that meets the Ministry of Education norms, and another group that has not yet finished acquiring basic reading skills. They still have language problems. When I was at the teachers college, they told me that the end of 1st grade means reading! But it doesn't work that way here. We try to adapt ourselves to each student through the end of 6th grade, so that they will be successful in the future at the best junior-high schools in Tel Aviv.

> Another Teacher
>
> At first I found things very difficult, but now I flow with the changes. We follow the Ministry of Education curriculum, but not blindly. Unfortunately, people outside the school do not always understand our unique problems and character.

In my discussions with the teachers, I found that many of them echoed Amira's educational and humanistic ideology. Her influence is evident in the way ideas are verbally expressed, as well as in the use of stories and the emphasis on narratives. The school culture has clearly been profoundly influenced by her leadership. As a cultural leader, her stories combined autobiographical comments with human observations drawn on professional and theoretical knowledge (Gibton, 2001). Many of the teachers have adopted a similar style of story-telling. The professional discourse of Amira and the teachers is rich in personal narratives and in examples from the field.

> Teacher (First Year in the School)
>
> Amira was very welcoming. "This is your home," she told me. I didn't feel that way in other places. There is also a sentimental side to the connection, because I come from the neighborhood myself. I was born in Shapira neighborhood, just a kilometer away. The surroundings have not changed that much. My experiences at school were similar to those of the students here. I am over forty, and I lived here from the age of seven to ten. Even then immigrants

came into the neighborhood. The experience was certainly different, but some aspects are very similar. There is a range of cultures here alongside the dominant, strong Western culture. I feel that they have the same experiences I did—the effort to cope with the Western world from a position of feelings of inferiority. I can say these things. I feel their lack of understanding and knowledge faced with the Western world. They suffer from feelings of inferiority despite the fact that they have their own rich and interesting cultures.

The School and the Curriculum

Just as Neve Shalom School has been obliged to cope without a curriculum adapted to the needs of a school devoted to coexistence, so must Bialik School confront the absence of a curriculum designed for multicultural schools.

Many of the questions addressed by the school relate to the field of socialization. The public education system in Israel strives—with varying levels of success—to "educate" citizens to identify with the state and with the Zionist ideal, to support the idea of democracy, and to gain an acquaintance with the canonical texts of the Jewish people. The Bible is viewed as the book that binds the Jewish people to its heritage—not necessarily a book related to faith, but one that raises the questions, themes, and dilemmas that characterize an emerging society. The Bible is a central subject in the Ministry of Education's curriculum, and the proposal to abolish the matriculation examination in this subject met with an uproar.

At Bialik School, the study of canonical texts such as the Bible has undergone a revolution. The reason for this is not a sudden passion for postmodernist insights among the teachers, but rather an intuitive realization that the "regular" pattern of Bible studies is inappropriate for a school that includes so many traditions, some of which are mutually contradictory.

Diary

Teachers' Comments

- There is a wide range of cultures here. You cannot teach them Bible on its own, you are thrown into other cultures. In history and literature, too, there are always broader ramifications.
- In geography lessons, every student learns about the countries and cultures of their classmates. The learning process is mutual.

• I draw the main ideas from the textbooks. I compare the Bible to the lost Mayan culture—there are many children here from Latin America. I showed them a film that raised interesting motifs relating to peoples that have disappeared from human history—motifs that we also find in our culture, such as cruelty. A culture of blood. In the Latin American cultures they sacrificed prisoners and so on. The Bible is also full of bloody stories. I wonder about which bits I should censor . . . I combine history with Bible studies. At first they do not understand what is going on, but eventually they get it. From Greece and Rome I choose themes relating to democracy. I also learn and teach about Far Eastern cultures when necessary. This is how the research projects are developed.

• How do we teach Torah? We do not always stick to the story, but look for universal motifs that are also found in other religions. I teach about the Creation stories of all the religions, including the scientific approach. In this way, every individual can build their own corner, even if they don't accept the usual way. Then the children go home and tell their parents what they learned at school, and ask to hear stories from their own faith, so that there is also an element of dialogue within the family.

• A senior representative of the National Religious Party visited the school and asked how we teach Torah. I told him that we present the Torah as a universal book. If we can't be a light to the nations, who can? I did not get the impression that he supports this approach.

In the regular education system there are no barriers preventing teachers from using their imagination and creativity in order to make the texts more relevant and interesting. At Bialik School, though, the teachers have no alternative. They find themselves in a situation in which they cannot function as agents of socialization who seek to replicate "good Jewish citizens," and accordingly they use their knowledge of other cultures to transform Bible studies into the comparative study of cultural myths. From agents of socialization they thus become cultural mediators.

Their function as cultural mediators leads them to examine the school subjects from a more interdisciplinary perspective than is usual in Israeli schools. The teachers realize that they are teaching "great stories." The multiplicity of cultures represented in the school leads them to develop a relativistic approach to culture in general, and to Israeli Jewish culture in particular. This creates difficulties among those who hold more particularistic and iconographic positions, as in the case of the representative of the National Religious Party mentioned above.

The multicultural, comparative, and interdisciplinary approach encourages teachers to develop an individual learning style among the students by having them prepare research projects. These projects enable the student to gain a deeper understanding of a given area, by not confining the research to a single field of knowledge. Sometimes, several students work together on a joint project addressing a central theme.

> Teacher
>
> We are preparing a curriculum that combines subjects—integrative or constructivistic studies, as they call it. The subject is built through the learning. We don't know where we will get to by the end of the year—we only know where we start. We create mixed-age study groups and we also pay attention to the children's interests. One group examines a given system—it might be a human system or the solar system—and another group examines interpersonal relations, while a third group examines the connection between life and art. All these ideas were raised by the students. The products are the subject of research, or an individual project, or even a report based on reading. From the 1st grade on the children can print their own stories at the school's computer laboratory.

Relative to the situation in regular schools, the statement "we don't know where we will get to by the end of the year" is quite radical. The tacit assumption is that if they do not know where they will get to by the end of the year, then the teachers do not have a fixed curriculum that they must follow. The teacher in the school does not serve as a conduit for conveying knowledge, and the student does not serve as a vessel for the receipt of knowledge. Learning takes place in a structured manner, and is developed through interaction among the different participants in the process; hierarchical relations exist between these participants, yet they both influence and are influenced by one another. This is a form of "cybernetic" knowledge that is characterized by the integration of both the individual and the subjective interpretations of all the participants. Adaptation takes place among the participants through the value-based process of negotiation that takes place by means of discourse, though there will never be a complete tessellation. In other words, knowledge has many faces (Keiny, 2002), and the school allows negotiation around conceptions of knowledge and reality, it does so from a position of cultural mediation and in a spirit of tolerance toward the other.

A further characteristic of the research groups at Bialik School is that they are mixed-age. This is unusual in ordinary schools, though it is encountered in alternative schools, such as Sudbury Valley School, Summerhill,[5] and at the Democratic School in Hadera, as we shall see below. Mixed-age groups also operate successfully in the social aspects of life at the school. Amira notes: "We allow many mixed-age activities in the formal system so that they know how to behave at recess and during their free time. If it works in the neighborhood, why shouldn't it work here?"

Self-Esteem and Culture

The teachers encourage the children in their research projects and in the learning process in general. The teachers in the school take every opportunity to help the students with every task, which gives them a sense of importance. The children have an advantage in the cultural realm. An ethnocentric approach would have made these children ashamed of their own culture and heritage and forced them to adopt the dominant culture. The only alternative for the children in such a situation would be to cling to their original culture and antagonistically reject the hegemonic culture. During its early years, the State of Israel attempted to create "new Israelis" who acted and behaved in keeping with the secular Zionist ethos of Socialist European Jews.[6] In recent years, Israeli society in general, and the educational world in particular, has seen an awakening of groups that feel that their voice was overlooked in the socialization process that accompanied the Western Zionist ethos. Neve Shalom School, Kedma School (see below), and the education system established by Shas[7] could all be considered manifestations of the rejection of hegemonic culture and the search for an alternative voice.

At Bialik School, the demographic change in the student population and the ideological change among the teachers constructed the multicultural ethos and the special quality of contact between teachers and students. It goes without saying that these students cannot and will not undergo a process of socialization to Zionism or Israeliness, because a national Israeli conscience divorced from Judaism has not yet emerged. Accordingly, each student is proud of his/her own heritage. The teachers respect each heritage and thank the students for the opportunity they present to learn about their culture. At the school no culture is perceived as superior.

The linguistic field is another source of pride. Teachers often encounter difficulties in explaining a particular point in Hebrew, since some students have still not reached a sufficient level of skill in the language. In such cases, the teachers call on one of the students to serve as an interpreter, looking for analogous or identical word in Russian, English, Spanish, or any other language. The children are very proud to be able to "teach the teacher," and enjoy playing an active role in the common learning process.

> Diary
>
> The teacher hands out worksheets. The students divide into small groups to discuss the concept of time. One of the children does not understand the concept, and the teacher encourages a girl in the class to explain it in his language. The worksheet relates to geographical concepts. After a while, the boy suddenly gets the idea, and excitedly begins to tell the teacher about his homeland. The girl translates, and the teacher shares the story with the whole class. The boy seems slightly embarrassed at the sudden attention, but remains happy and smiling until the end of the lesson. After the lesson, he tells me in Spanish that he was surprised that "the other children and the teacher found it all interesting." He has been in Israel for two months. His mother does not have a residency permit—they arrived as tourists. Before coming to Israel, she worked for a while in Cairo.

We sit in Amira's office and I give her my impression of B, an Argentinean girl who is studying in the sixth grade. B has an impressive ability to express herself in Spanish and Hebrew. She wants to study in Municipal School "A" because of her talent for dancing and singing. I ask Amira what will happen when the girl reaches the twelfth grade.

> Amira
>
> She may not be able to take the matriculation examinations, because she does not have any identifying papers . . . She will not be able to take part in training for National Service, or to get a driver's license or to join the army with the rest of her class. When she returns to her country of origin, she will face problems, because her Spanish will not be very good (she has been in Israel since 1st grade). She will have problems studying at university there, too. Israel should follow the example of Holland, Germany

and other countries with high levels of immigration, and find a way to naturalize the immigrants. What happens in Israel is that from the age of eighteen, this girl will have the same status as her mother—an illegal alien in the State of Israel. The only way to naturalize in Israel is under the Law of Return. There is no other way to belong to Israeli society. In other countries immigrants can be naturalized, and once they are citizens they have a sense of belonging. In Israel, even if you are a citizen you cannot belong unless you are Jewish.

Question

What about the Druze? What are the chances for integrating foreigners in Israeli society?

Amira

The Druze marry among themselves and this is their focus of identification. A Druze man who marries a Jewish woman could convert, but he will lose his connections with his own society and religion. Foreigners cannot belong to Israeli society. Israel has still not paid attention to this problem. If there are 300,000 statusless migrant laborers in Israel, they should be naturalized. But then they would be able to vote for the Knesset, and the number of mandates they could control would give them considerable power. The parents would also be able to influence the curriculum in schools. At the moment, we teach as we see best. But what will happen in the future? We have to begin to prepare. I do not know what the best solution is, but we need to think about it and plan for the future. We have a responsibility to develop binding policy addressing these problems.

In Another Interview

The children have problems with their lack of identity, rootlessness and instability. We deal with these problems as best as we can by developing a sense of belonging to this place [the school]. It is all a product of migration. The children who were born in Israel to foreign residents receive a birth certificate without an identity number, but they feel that they belong here. There is a problem in Israel that people cannot belong to the collective. We have to ask ourselves these questions, tough as they are. We cannot drop the subject—it's an urgent matter.

The school nurtures the self-esteem of students by respecting their culture and language and by teaching everyone a few words of that language. Above all, the children play a proactive role in the joint learning process through the different study groups in the school. It should be recalled, however, that this policy was developed on an ad hoc basis, without prior planning. The demographic problem was created before the school developed a solution for it and the ideology behind it. Amira's comments show the complexities involved in the issue of migration, particularly as someone who grew up in the heartland of the "old" Zionism. She does not offer unambiguous or simplistic solutions. She underscores the difficulties faced by non-Jewish immigrants in integrating into a society that still maintains a direct connection between Judaism and nationality. She sharpens the discussion by giving the example of the South Lebanese Army families and Arab collaborators,[8] whose status is the most problematic.[9] On the major question of the tension between Israel as a Jewish state and Israel as the state of all its citizens, Amira's position seems relatively tentative. This in no way diminishes her devotion to the children. To paraphrase, she seems to be saying, "I did not create the problem of the migrant laborers, but the problem exists, and since I am responsible for the system that cares for them, I will do all I can to see that they get everything they deserve." Amira takes responsibility for what is happening here and now, leaving to others in Israeli society the discussion on immigration and its ramifications. Thus she adopts a vague position on the question of migration, while showing unswerving devotion to the immigrants as people—the exact opposite of the way in which Israeli society addresses Jewish immigration, particularly in the case of Russian Jews. Amira, with her Socialist values, acts in accordance with humanistic values of human rights in general, and children's rights in particular. If the State of Israel follows prevailing demographic trends in the Western world, it may be assumed that she will be able to continue to work in this field for many years to come.

The school works hard to give the children and their parents a sense of belonging. The parents receive an explanatory booklet about the school, translated into Russian, Spanish, Arabic, and other languages. The translations were prepared by the teachers themselves, taking into account cultural needs. For example, it emerged that in Russia it is not customary for teachers to write notes to the parents about lice, so this subject was omitted from the Russian version of the booklet.

One of the teachers who prepared the translation immigrated to Israel several years ago. She recalls how her daughter struggled taking examinations in Hebrew, and how after having learnt the language she cries out of relief when she translates the booklets for parents, because she feels that she is closing a circle.

The Status of Immigrant Children and the Establishment: The School's Position

Diary

Amira reads the Convention on Children's Rights, which refers to the good of the child. Israel has a Compulsory Education Law and a National Health Law. If there is no correlation between the child's status and that of their parents, as demanded by the Convention and by the Compulsory Education Law, the children of immigrants should be enabled to benefit from these laws, despite their parents' status (illegal aliens). This is the situation as required by the distinction between the status of children and the status of parents.

"That's what the situation is in our country," Amira concludes.

The relative ease with which people can move from one country to another in search of better living conditions, adventure, work, professional training, and so on means that many people now experience cross-cultural encounters. Cross-cultural contact does not automatically lead to greater understanding of the other.[10] Both sides need to become acquainted with one another from a position of equality. In situations of cross-cultural encounters between a hegemonic culture and a traditional culture, the institutions tend to reinforce the mechanisms of inequality between the different groups. Moreover, the members of the dominant group tend, sometimes unconsciously, to believe that the values of the hegemonic culture are the best and the most efficient for the entire population.[11] Dissatisfaction with the inadequate stress on integration in the Israeli education system may be due in part to the sense that ultimately only one group was represented in educational ideology. The fact that all the children at Bialik face problems mutes the identification of a hegemonic or dominant group. All the students at the school need some form of institutional assistance, and the institution—the teachers and the ancillary staff—do their best to help without attempting to impose cultural coercion or conversion. It is perhaps easier for the teachers to take this approach since it is uncertain whether the children will remain in Israel. However, the encounter itself and the effort to adapt the curriculum to a multicultural society clearly lead the teachers to examine their attitudes toward their own identity, their profession, and various social questions raised by this encounter.

At Bialik School, an encounter takes place between a group of teachers that belongs to the elite stratum of Israeli society and identifies ideologically with the state, and a group of students whose identification is functional. There is no aspiration or pretension to convert their functional identification into ideological identification ("whitening," as Amira puts it), since Israeli society has not yet developed a civic Israeli identity. Yet the presence of this educational framework within the first Hebrew city and the increase in functional immigration to Israel[12] further underscore the need to discuss these issues.

A basic sociological insight argues that bureaucratic systems potentially create evil. Bialik School constantly encounters evidence of the ways in which the system employs codes that literally and figuratively "delete" these students. The Ministry of Education records use the severe heading "error code" to relate to the children at the school. In the future, they will be unable to take matriculation examinations, because their identity number lines carry the legend "000000000" in place of an identity number. Despite the commitment of the State of Israel to children, and despite the fact that Israel has signed the Convention on the Rights of the Child, these children do not have the right to medical treatment in the Israeli health system. The staff at the school undertake unusual risks in order to ensure that these students experience elements of normative life of Israeli school students, such as annual outings. As noted above, many teachers at the school take on a quasi-parental role, due not to overinvolvement but rather to a moral outlook that places the needs of the child at the center.[13]

Management Style

The atmosphere at the school is profoundly influenced by Amira's management style. The teachers are aware that she practices what she preaches and avoids pompous slogans and clichés. She clarifies her position in simple, understandable, and communicative ways and provides a personal example. Her attitude to other systems is complex. She has developed very fruitful relationships that help her advance the goals of the school, but she is not afraid to criticize when she is dissatisfied with the functioning of particular systems, particularly the Ministry of Education, of which she has very high expectations.

Meeting with a Senior Representative of an Agency (not the Ministry of Education) that Assists the School

> Amira
>
> I am very disappointed with the Ministry of Education headquarters, and it doesn't have anything to do with a particular minister. The current situation shows that over through the years since Israel was established, we have not managed to narrow gaps. There is a very clear division between "nurtured" areas and unsuccessful areas. Instead of addressing these gaps with courage and determination, the approach has been piecemeal. They send out tests and develop programs. Unfortunately, the system is constantly preoccupied with its failings. But in all the places that are considered failures, every so often a successful model is found. Maybe they should try to find out why someone can succeed where everyone else has failed? After all, "needs support" is a socioeconomic label, not a genetic disease. After fifty years of independence, the label will become an incurable disease if it isn't dealt with properly. The problem doesn't rest entirely with the teachers and principals. The headquarters of the Ministry of Education needs to understand that its first commitment should be to weaker places.

Amira's criticism is the result of an attempt to create uniform standards for human systems that are inherently unequal. Shderot cannot be treated in the same way as North Tel Aviv. Amira's comments relate mainly to the centralization in the education system and the lack of consideration for context that is of utmost importance in understanding the needs of the child in the system. She feels, perhaps with justification, that Bialik School is an example of what can be achieved with a proper mapping of the educational context, the needs of children in general, and the needs of each individual child:

> The standard test of the Ministry of Education should examine them in their own language. We need to examine whether they are sleeping at night. Should there be questions in the examinations about taking money out from the ATM and buying a ticket to the theater? Don't the people in the Ministry of Education know that fifty percent of the children in Israel do not take money from the

> ATM and go to the theater? They can keep on sending the reading comprehension tests, but I told them that enough is enough. They exempted me because I am considered an experimental principal.

> Teacher (third Grade)
>
> We came to the conclusion that we cannot work according to the standard approach of the Ministry of Education. We adapted ourselves to meet the children's needs. There are children here who do not know how to read, and at the moment they are having reinforcement classes. At a regular Israeli school, children of this age are already reading. In our group we have children who speak Spanish, Turkish and Romanian. We try to bring everyone together and teach them Hebrew.

According to Amira's educational philosophy, it is the child who leads the process. Children can be motivated to learn and to enjoy learning, provided they enjoy the conditions in school that facilitate the educational process. Amira is aware that numerous systems exist to serve children's needs, and, as far as possible, she tries to arrange for such systems to intervene to help the children at the school. She has managed to recruit many and win their cooperation by emphasizing the need to help children in need. Amira calls the school "Israel's fig leaf." The school is visited by a large number of delegations organized by the Ministry of Foreign Affairs, Tel Aviv Municipality, and the Ministry of Education. The delegations come to see the school's success and learn from it.

Amira believes that the school is not merely a successful solution to a local problem, but also a model for other schools that must be established by the State of Israel if it wishes to meet the needs of mixed populations in different places. Amira is convinced that Bialik School constitutes a success story in dealing with "difficult" populations in general, particularly given the statistics of failure that are presented and indeed perpetuated every year by the standard tests.

Amira accepts a heavy responsibility and persuades the teachers at the school to share in this responsibility. She does not hesitate to change the curriculum when she feels that it is failing to meet the children's needs. She is sharply critical of the Ministry of Education's standard tests, which ignore the educational context in matters of content and language. She suggests, on the basis of the educational theory of intercultural encounters, that these tests are culturally biased.

Amira feels that the Ministry of Education often lags behind when it comes to thinking progressively about schools. She is critical of its conservative approach to education in general, particularly of its confused and contradictory attitude to the children of migrant laborers.

> Amira
>
> Many principals are ahead of the Ministry of Education. They move teachers around, change the scheduling of lessons and the structure of classes, and choose themes that are appropriate for them. For example, mathematics teaching may be taken from class educators and given to teachers who were trained to teach the subject. Sometimes the principals turn the school into a kind of community center offering all kinds of activities, because they appreciate that socialization is one of the main problems that needs addressing today. This is also part of the school's mandate. Yossi Sarid [a former education minister] said that he could not even move a nail in the ministry. Dismissing a single employee leads to political riot. Yitzhak Levy was glad to move to another ministry—he told me that at least in the Ministry of Housing you have a chance to do something. They have a department there for overcoming obstacles. If you want to build a new neighborhood, you have to overcome all kinds of objections. The Ministry of Housing has an expert in every field. First they deal with the objections, and then they start planning. But in the Ministry of Education, they "build the neighborhood" on paper, send out the programs, see that it isn't working, start building something new, and so on. Today, the principals are largely autonomous. They know what is best for the communities they serve. People in the field should be given a greater measure of independence to develop programs and areas of content, without being subject to coercion from above.

In her management style, Amira recalls the "wild" style of management, as Gibton termed the work of a group of Israeli principals working in autonomous schools. Adapting Sergiovanni's theory, Gibton identifies among the principals he studied characteristics similar to those Amira displays: criticism of the system within which the school operates, mastery of the terminology of educational research combined with an ability to identify discrepancies between theory and practice, and emphasis on illogical and problematic aspects of the situation in the field. Wild principals "can display leadership inside the school and elsewhere in order to lead the educational institution in difficult times."[14] However, Amira opposes the concept of autonomous schools (in the financial sense of the term), since she feels this will exacerbate social gaps and perpetuate inequality. On this subject she finds allies in Kedma School, as we see below.[15]

Empowerment

Although Amira is the dominant figure at Bialik School, she works hard to empower the other participants within the school community, from the newest student to the most senior teacher. The school ethos is so well established that it seems that the school would continue to operate in a similar manner even if Amira were not physically there. Amira's work with Anat, who is now responsible for all the training activities in the school, is particularly impressive. Anat comes to the school almost every day in order to help the teachers as a mediator between the students' culture and that of the teachers. Her help extends to almost every possible human sphere.

> Diary
>
> There is a lot of noise in the staffroom. I try to listen to a conversation between Anat and two teachers. She patiently explains how to map abilities in the classroom. I cannot hear everything, just fragments of sentences: "That way you can see what the two or three weakest skills are in this class;" or 'you should order the tests according to the number of mistakes, not the grades," or "for example, that girl didn't answer the question not because of a lack of skills, but because she didn't understand the test." The teachers' body language suggests that they are relaxed and attentive.

Anat is extremely grateful to Amira for the opportunity she has been given to develop and grow within the school. Amira recalls that she had to "drag" Anat from her third grade class at the end of the corridor and persuade her to give up much of her private life in order to develop the management potential she saw in her. Anat is now completing her MA studies at Bar Ilan University, and Amira believes that she should go on to PhD studies and become a lecturer. Amira says that Anat is an expert in the field of developing literacy skills among children. She provides training for all the staff in the school and has a natural talent for leadership.

Anat refers to her years in the school as a period of "rebirth." From the teachers' perspective, it is Amira who sets the tone while Anat puts theory into practice. My own impression is that each one of them, as well as several other teachers, could, when the time comes, take on the task of management in both the practical and the ideological spheres.

Anat

We change all the time as the needs of the school change. The teachers are flexible and professional, and most of them are highly motivated. We are part of the Tel Aviv District, but we set our own borders. We take elements from programs that are suitable for us, recognizing that each school is different—and we are particularly so. Since we have our own special problems, it is considered legitimate that we work differently. I adapt the outlines for each teacher and each class. I meet teachers through the course of the day, and through each meeting I change a little. The teachers get what they need. I don't impose things, I don't get caught up in an ego trip and I don't whine about the difficulties. I always try to look for the positive, and show that every problem has a solution—we just need to look for it.

For example, a new teacher came to the school, and found herself in a 3rd grade class with a lower level than she had expected. Now, two weeks later, the situation in the class is a bit better. But yesterday she felt that she was on the verge of giving up. I gave her emergency support, including emotional support. Sometimes close contact and bonds can provide enough support to let the teacher continue their work—that was what happened yesterday.

Another issue is how I divide my time. I try to free myself up when a new problem arises—I put everything else aside and try to give the feeling that I'm there to provide advice and help. I don't hold a gun to the teacher's head and force them to change, put try to provide guidance and reach a higher level each time.

The Ministry of Education has very clear policy on the subject of language. There is a defined range for each age group. We do not follow their curricula, but develop our own adapted activities, which are not always at the level demanded by the ministry. The reality here is very heterogeneous. The child is seen as leading the process- not as a slogan, but as a simple fact of life. Throughout the week we provide language skills in small groups. Most of the teachers are already used to working this way.

I am a pedagogic instructor in the school. My specialization is in the field of language skills. All my time is devoted to training—some hours are fixed and others are flexible. I am also responsible

> for the field of literacy, and I am in contact with the Ministry of Education on this field. The contacts with the ministry or the municipality depend on the person I am working with. We explain the rationale behind our approach and receive assistance. They know us by now. This way of working means that I have to keep learning and keep myself informed. We use all kinds of tools for formative evaluation. Once a week we come together for brainstorming to discuss what's happening. Two new teachers who came to the school were amazed at how we work compared with their previous schools. That's a good sign. Amira and I are so preoccupied with our work that we cannot always see what is happening here. Teachers who come here admire the calm atmosphere and the attitude toward the children, and this is also a form of evaluation.

The empowerment of the students in the school is no less impressive than that of the teachers. Due to the complex needs of those who learn in the school, the system has become highly flexible. In a manner that is reminiscent of university education, an individual program or timetable is developed for each child.

Amira and Some Teachers Talk to Students Visiting the School

One of the Bialik teachers say, "The child takes responsibility for their own learning. For example, one of the children chose a study project in which he had to define the concept of limits. We ask him why he chose this subject and discuss the process with him."

Another teacher adds, "The school is subject to the 'Shetrit Law' (which established a 41-hour school week). Accordingly, the children spend eight hours a day at school, except for Friday which is a short day. Some of the learning takes place in the community center, which offers activity groups such as drama and judo. The groups are not divided according to classes, but according to areas of interest."

At this point, a visitor asks if the children are not asked to take on an unreasonable level of responsibility. The teachers reply that they do not think so, noting that each child is individually monitored. They believe that "the more you make someone accept significant and real responsibility, the more they will grow."

Amira adds, "This is one of the problems of the education system. In our school, a child who is wandering around the corridor on their own feels uncomfortable, because all the other children have found a place in one of the classes. It is hard to get used to this kind of system in the first month or two, but if they are patient they adapt to the demands. This is a way of life, not a matter of luxury. We are used to integrating our emotions in our work. Someone who does not love children cannot work here."

Amira

We have plenty of opportunities to create mixed-age models. Three hundred children cannot go out into the yard on a rainy day, so they stay here and play in the corridor or in the classes. If they haven't been taught how to behave and play in mixed-age groups, what will happen in the corridor? The school has undertaken a serious goal of using every opportunity to coach the children in how to function in a heterogeneous society. We have this opportunity because they don't really make demands of us, because after all we are educating children who "require nurturing," or because we work with the children of migrant laborers, so how serious can it be? Instead of enjoying this lack of expectations, we developed other areas. We don't have to make them into standard Israelis, we can relate to their culture.

When someone gets stuck on an iceberg, they have to take the decision to act. An iceberg can be a problem, but it can also become an opportunity. The same is true of diversity. This kind of multiculturalism and poverty can be a basis for progress or it can turn into a quagmire. Our success is that Israel's "fig leaf" has been turned into an advantage. The State of Israel cannot allow its "fig leaf" to be anything other than presentable. Tel Aviv Municipality has seen to that. We have National Service volunteers, teacher-soldiers and welfare workers. One of the questions is what you do with the budgets. I thought it was important to establish a strong support team that included pedagogic instructors. These are teachers from within the school who take on special tasks. I make sure that hours are available for them to train the staff and that they are rewarded for their work. My assumption is that an hour's training for two teachers will be beneficial for forty students. If the same hour is devoted to working with one child, then we have only helped one individual. That's why we invest more time in staff training. Even if the approach of diverting hours for training would be criticized by the state, it doesn't bother me. Over the past two months the Minister of Education has asked us to submit programs for establishing additional schools like ours. We have our own truth and we're sticking to it. We hope we can persuade many others to follow the same path.

Amira Talks to the Students

The contacts with each parent take place over the telephone and through letters in a relevant language. Most of the parents work from morning to night and it is very hard to arrange meetings, so we have to hold meetings at unusual hours. Most of the parents are very interested in what is happening with their children, but they express it differently. At other schools they would say "the parent didn't come to the meeting because they aren't interested." But our parents work from seven in the morning through seven at night, or even ten o'clock at night, and yet they come to a meeting at ten o'clock. That would not happen at other schools. We have a general meeting twice a year, when parents can come and meet with us at any time from morning until nine o'clock in the evening. There are also individual meetings with the teachers.

Students: Don't you find it hard?

Amira: No, it isn't hard. Life isn't hard when you're really doing something. It's a way of life for us, a thought process and behavior pattern that are different from everything we have known before. I am sure that if this approach was adopted throughout Israel, Israeli children would feel better.

Amira

In this school, the children always like the teachers. That made me think about something. Usually, children hate school. I don't know why. Why do they like the Scouts or other youth movements or the community center? Then one day I realized that the children here need love. We can't work without it. We began by creating a sense of personal commitment among the teachers, and this led to emotional bonds. I don't if you can call this parental love, but it is certainly love that the children need and want. Commitment to love can lead to commitment to academic, social and educational results.

We try to give the children tools so that they can help themselves. They need to take part in planning their own day from an early age to prepare them for adult life. The children send most of the day in small study groups based on their areas of interest and their academic level. Every student has an individual study track. Children who share a field of interest come together at a given time. For example, in the afternoon a mixed group goes to study at the community center, where they divide into four activity classes. In this way they learn to consume services in the community. The home base class, just like many other concepts, is not treated as a sacred cow in this school. Any principle or subject can be changed as necessary, on the basis of a structured rationale and by means of a process that is subject to evaluation.

Diary

R, a teacher at the school, is getting married. The students are very excited, and the teachers report that they worked very hard to buy her a present. The students also wrote very moving letters. They have an unusual sense of gratitude. In one case of severe hardship, the teachers managed to provide shoes, clothes and household equipment for a family.

Diary

I sit in the school office. There is a lot of noise outside. A, a 4th grade student from Colombia, has been injured. They are trying to contact her mother, who works as a cleaner and has a mobile phone. Amira and some of the teachers sit nearby. Nobody seems agitated or pressured. A tells me that she has been in Israel for a year and a half and likes it. Amira is concerned that A's mother will have to pay a lot of money for medical treatment. Accordingly, she tries to give the children basic medical attention herself, such as putting on a bandage. Some children are brought into Amira's office to explain what happened. Given the dramatic event, they give their versions relatively calmly. Amira tells the boy most closely involved in the incident: "You're a big boy, we shouldn't have to watch out for our all the time. You need to take responsibility for your actions."

Amira asks me to translate her comments, since the boy does not speak very good Hebrew. I emphasize the financial ramifications of the girl's injuries. "Very bad news," his teacher adds in pidgin Spanish, "every time this happens it makes problems for your parents." I talk to the boy. It emerges that, like many students at the school, he is in Israel with just one of his parents. He likes the class and his teacher. I also talk with a friend of his, a boy from Ecuador. He also knows Arabic, because his parents lived in Cairo for a while. Amira asked me to find out where he comes to school from. She sense that he may not be telling the truth about his home situation. She thinks that he comes from Beit Dagan [outside Tel Aviv], in which case she cannot accept him at the school. His answers are vague.

After everyone calms down, Amira shows me a booklet presenting the school in several languages. The booklet includes the telephone numbers of social workers and ancillary staff, important information, the rules of the school and reference to Israeli law. The idea is to explain differences between Israeli culture and the students' original cultures.

Amira: "I explain to them that I have an obligation to report certain things to the authorities. For example, one father told me that he would beat his child to stop him misbehaving. I warned him that this was not allowed and I would have to report him, which I don't want to do.' Amira emphasizes that it is important to explain the Regular Attendance Law to the parents—the law requires students in Israel to attend school regularly. If the parents do not understand, the teachers explain again and again in their own language.

Amira tells me that when a child does not understand something, she tells them to sit down and think about it. She adds that soon they will talk to the students about elections, since general elections will be held shortly.

Amira
The education system is always preoccupied with mediocrity and concentrates on failures, instead of identifying successful schools, examining how they work and learning from them. The world doesn't work like that, it works like a mirror. Look at the hi-tech companies. The education system also needs to look for successful models and learn from them, instead of talking obsessively about failures.

Conclusion

Tel Aviv, the first Hebrew-speaking city, lies on the Mediterranean coast. It is home to the international Bialik School. The school is international, although no one has ever declared it to be such. Nearby, at least in geographical terms, lies the International School, which serves the children of diplomats, is very expensive, and teaches in English. The word "international" brings to mind associations of specific socioeconomic classes. Along with the entire area of south Tel Aviv, Bialik School has undergone a transformation in recent years. The city that always longed to be cosmopolitan now includes neighborhoods that are home to poorly paid migrant laborers who have come to make a living. These workers come to Tel Aviv not out of ideological identification, but as part of a functional migration designed to meet basic needs. The State of Israel does not have any clear policy on the subject of migrant laborers. The approach is short-term and crisis-oriented, without devoting time to in-depth discussion of the ramifications of creating such clearly distinct communities within city centers and other locations in Israel.

Government policy toward migrant laborers is inconsistent and even contradictory. Under Amira's inspired leadership, Bialik School tries to meet the needs of the children of migrant laborers, children who grow up in squalid and overcrowded conditions in a society that accepts them only because of the functional tasks they perform. Israel does not have any policy on how to treat these children. Amira is very familiar with the laws and regulations and has discovered that the official laws, regulations, and policies are sometimes contradictory. In the face of these contradictions, she is required to take decisions based on the universal principles that have been adopted by the State of Israel, as embodied in the Basic Law: Human Dignity and Liberty, the Convention on the Rights of the Child, and other similar laws. She is sharply critical of the policy of deporting migrant laborers and of the fact that the children cannot continue their studies, take the matriculation examinations, or go to university. Amira does not have an unequivocal answer when asked what Israel should do in this field. As long as the problem exists, however, she will strive to give the children support and help, based on a sense of universal human solidarity, respect for human dignity, and her own Socialist upbringing. The fact that one of her own sons has decided to establish a family with a non-Jewish woman raises various questions and associations for her. Amira often looks at the problems in the school through the prism of her own personal experiences.

All You Need Is Love

Bialik School underwent a transformation due to demographic changes and was not set up deliberately as an international school. As a result, it has been forced retrospectively to adopt an ideology. John Lennon's song seems to sum up the ethos of this school, which is based on an Ideology of Repair through love. In an storm of contradictions, difficulties, and complexities raised by encounters with the other (the foreign children), the staff declines to make value judgments, instead they repeat the mantra of love as they search for the common and unifying elements of human experience. It is no coincidence that the school's motto reads "The teachers at this school love children." The teachers at the school develop sensitivity to issues of multiculturalism and learn to diversify their teaching methods in order to enable each child to make the most of the educational experience. Teachers learn to understand diversity and to support the children by taking care of even those basic needs that in the regular system are met by the family or other community institutions.

The teachers themselves are still undergoing a transformation. The school illustrates what educational leadership can achieve when it provides teachers with a clear ideology, an understandable direction, and good guidance. The ideology of the school staff is not ethereal or remote, but rather a practical ideology of Repair. All those involved in the process are expected to form part of the community of Repair. The school makes extensive use of case studies in order to understand problems in the practical field and to improve educational work.

Through a familiarization with the surrounding systems, and thanks to strong communication skills, the leadership of the school has managed to develop contacts and partnerships that meet the different needs of the students, thus helping both the children and their families. The school also makes good use of its status as a school "requiring nurturing" situated in a disadvantaged neighborhood.

Bialik School is a living challenge to sociological predictions that expect a school of this kind to be marred by delinquency, violence, alienation, and poor achievements. The school manages to help children who in any other situation would be labeled as being at high risk. The teachers believe in the children's capacity to learn in an atmosphere that is warm, responsive to their needs, and supportive, particularly when there is an understanding of multicultural issues. Bialik School is an island of tolerance in a sea of alienation, insensitivity, and bureaucratic evil. The school shows the tremendous potential inherent in an education system to deal with unusual challenges. The success of Bialik School has ramifications even for schools facing a less threatening level of risk, schools such as those in development towns with large concentrations of residents of "low" socioeconomic status. If it works here, why should it not work there?

Although the staff at the school is largely preoccupied with the here-and-now, it has developed sensitivity, tolerance, and humanism in a universalistic sense of the terms. The constant quest for Repair and the efforts to ease the suffering of others lead to processes of profound introspection and critical examination of the unwieldy and inflexible Israeli unsuitable bureaucratic systems that implement contradictory policies and adopt approaches to the phenomenon of the migrant laborers. The growth of functional migration to Israel, assuming that many of the migrants will ultimately remain in the country, will encourage the development of a civil society.

Policies and ideologies dictated from above are not always consonant with ground realities. When a social problem emerges, as in the case of Bialik School, mechanisms are developed that offer both practical and ideological solutions. Bialik School offers a model of an Israeli educator who has a universalistic commitment to all children, an educators who hopes that his/her humanistic policy will have an effect on the education system as a whole and on other spheres of public life. A hope that if, realized, will help the State of Israel begin to provide a proper humanistic response to the needs of foreign residents and their children.

CHAPTER THREE

Keshet School

(Kindergarten through 10th grade, approximately 300 students, developing a senior-high section)[1]

Diary

I usually come to the school leisurely, riding my bike through the streets of the Katamonim neighborhood of Jerusalem. Keshet School moved to the neighborhood in its second year (the 2001/2 school year was the school's seventh year of operation). Real estate prices rose slightly in the neighborhood during a certain period as Yuppies moved in, attracted by the possibilities for expanding the small low-rise homes in the area. The neighborhood does not have a uniform appearance. Some of the buildings have the appearance of housing projects, while others boast extensions covered in Jerusalem stone. The majority of the population is still the original core of lower middle-class residents. Many homes bear evidence of support for the Mizrachi religious movement Shas and other religious emblems and amulets. Support for the Betar Jerusalem soccer team is also much in evidence.

Keshet School is housed in an old and unwieldy building in the heart of the Katamonim neighborhood. The school has an elitist image, though, as we shall see shortly, this is misleading. The school is committed to recruiting at least one-third of its students from the neighborhood, and the positive atmosphere among the students suggests that this process of integration is successful.

The school was established on the basis of an ideology of Repair that seeks to bridge, or at least narrow, the gulf between secular and religious Jews in the State of Israel. During the early years of the state, a number of key policy decisions led to the intensive involvement of religion in Israeli politics. One of the consequences of this process was the growth of anti-Orthodox sentiments among many sections of Israeli society, particularly among secular Jews who favor the ethos of civil and Western liberty as a role model. Even among the religious population there are many who oppose the involvement of religion in political life, feeling that this only serves to alienate many Israeli Jews who might be more open to Judaism were it not for what they perceive as coercion.

The rift between secular and religious Jews in Israel is manifested in various spheres. Much of the anger of secular Israelis is directed at the ultra-Orthodox population, which is perceived as sectarian, insular, selfish, and unwilling to accept the burdens of employment and military service. These emotions formed the backdrop for the development of the political party Shinui. Attitudes toward the State of Israel in the ultra-Orthodox community actually range from identification and acceptance (bringing them close to the positions of the national religious stream) to total rejection. The ultra-orthodox society shows signs of a sense of moral supremacy and believes that it is maintaining the one true form of Orthodox Judaism. Ultra-Orthodox Jews question the Jewish credentials of secular Israelis, whom they consider at best misled innocents or an uneducated rabble. The ultra-Orthodox community functions as a closed society and makes every effort to avoid exposing its children to the symbols, culture, and values of the secular world.

Contacts between secular and national religious Israelis are also limited in many cases. Children study in separate education systems and the two populations have differing opinions regarding the Jewish character of the state, the role of religion in government, and the question of "Who is a Jew?" In addition, differences emerge on broader political issues: the national religious stream is strongly identified with the establishment of Jewish settlements in Judea, Samaria, and the Gaza Strip. Many secular Israelis do not distinguish between the ultra-Orthodox religion the national religions, viewing all religious Jews as an obstacle to the emergence of a progressive and enlightened society in Israel.[2]

The gulf between the two populations is manifested in the lack of mutual cultural understanding that militates against positive contacts and rapprochement between secular and religious Jews.

Keshet School was established on the basis of an ideology that seeks to repair the secular-religious gulf in the State of Israel. Ruti Lehavi,[3] who conceived and established the school, often comments that Israel made a mistake in the 1950s when it decided to establish two separate education tracks, state and state-religious. She decided to establish a school that would "attract parents interested in a good education for their children," regardless of their religious or secular identity. It was important for her to define what constituted a "good" school; part of the school ethos was to be manifested in the possibility of reaching "a common definition that does not relate to our lifestyles, while bringing our lifestyles to the school."

Ruti developed the idea of the school while spending two years as a fellow in the Mandel Program for Educational Leadership. As in the case of Neve Shalom School, the assumption is that a properly guided encounter can help children from both sides to develop their identity while respecting and understanding the other. In the case of Keshet School, the "other" is a Jewish other, and ostensibly, therefore, the task might seem simpler and easier here than in Neve Shalom.

Ruti often emphasizes that the school is different from other frameworks that create short-term encounters.[4] She is concerned that some of these encounters actually perpetuate and exacerbate stereotypes. Keshet School aims to provide a response in everyday life over an extended period. Ruti appears to have developed the guiding principles behind the establishment of the school even before it opened:

> We are involved in guidance in the broadest sense of the term. The school was conceived three years before it opened, in an effort to think about all aspects before the opening. We considered various scenarios that might arise during the everyday work of the school. We considered a pre-constructed encounter, including the "half and half" ideology, which relates to the staff, children and teachers.

The goal of Keshet School is to strengthen the identities of the students from different groups, be it secular or religious. The school does not attempt to blur the distinctions, it recognizes them while learning together and developing tolerance for others. As we have seen, this approach is being put into practice in Neve Shalom School. At Keshet School, however, it transpires that the process is actually more complex. While in Neve Shalom there is no real possibility for students to change their identity, in Keshet School, at least in theory, every student might potentially abandon their identity in favor of that of the "other side." The school does not encourage this possibility, but the children enjoy freedom of thought and determination. Thus, while the school attempts to engage in a clarification and upholding of differences there are also, mechanisms that act to blur the separate identities and to blend the cultural motifs of both populations.[5]

It may be that one reason for the blurring of secular and religious identities that takes place in the school relates to the parents' decision to send their children to a school of this type. Many of the religious parents at the school do not seem to be "regular Orthodox" Israelis and do not belong to the mainstream of the national religious movement. Many of these parents have a strong commitment to democratic values, hold more left-wing political positions than is usual among religious Jews, are open to the secular world and have even learned to identify with it. These parents seek interaction with secular Jews as part of the socialization process of their children.

> Tova Avichai-Kramer, principal of the elementary school
>
> The diverse beliefs of the students are a challenge that must be addressed. Teachers must be prepared for this encounter and must welcome the challenge. The religious students come from democratic homes that are open to the secular world. Children from classic Orthodox homes would not come to this school.

As we have seen, it is not easy to be a teacher in a multicultural school. Here, however, the difficulty is even greater due to the declared goals of Keshet School. The children's identities are more complex than the possibly artificial dichotomy of religious versus secular.

Before coming to Keshet, Tova was principal of the school in Kfar Eddumim. This is a settlement in the West Bank populated by religious and secular Jews, and the children in the settlement study in a joint school that is one of the pioneers in bringing together both populations in a single school. While the religious children at the school say their prayers, the secular children address other issues (though the school does not offer the spiritual alternative that has been developed in Keshet School). There is no "spiritual" encounter designed to facilitate the search for secular identity. The school in Kfar Eddumim does not feel the need to strengthen secular identity. Moreover, the school in the settlement is a natural continuation of the surrounding community, whereas Keshet School is a city-wide school that strives to create a community around itself. Keshet School also strives to achieve symmetry in the choice of staff members. The school operates under the state system as an experimental school. In the school in Kfar Eddumim, by contrast, most of the staff are religious and the school is affiliated to the national-religious track.

Part of the difficulty in creating a distinct secular identity may be the result of the different approaches. While the religious approach is characterized by commitment to a life full of texts and unifying symbolism, the secular approach is characterized by the absence of any binding tradition; if texts are selected, they are not coercive, they may or may not be followed or experienced as individuals choose. The situation is therefore different from that in schools such as Neve Shalom or Bialik, where the emphasis is on encounters between different traditions, each of which has its characteristic ceremonies and myths. In these schools, the students are invited to "visit" the other tradition in order to learn more about the other, but with no attempt to create cultural conversion. The problem of the secular side in Keshet School, it seems, is the absence of a fixed system of ceremonies and traditions (plurality serves only to make it difficult to define such a system). On occasions, the school seems to "create" a secular ceremony with spiritual value and weight in order to "fill" the space created while the religious side is involved, for example, in morning prayers.

> Diary
>
> B, a teacher in the 5th grade, gathers together the secular children and reads a poem by Tchernichowsky. A short discussion develops on the poem, but the teacher quickly steers the students to a discussion on the subject "when do I behave the way I feel, and when do I behave the way society expects me to?" The discussion is inspired by the content of the poem. The children take part in the discussion without any special effort on the part of the teacher. They bring examples from their own experience, some of which are impressive given their young age. When one of them talks, the others listen. Occasionally a student interrupts, but rarely and briefly. The school is overcrowded and we sit in a circle in a small class situated in a bomb shelter.

The school in Kfar Eddumim mentioned by the principal may be more typical of the religious-secular encounter in Israel. Religious Jews are committed to a tradition that requires them to devote time to different rituals; it is certainly not easy to get up early in the morning to "do the Lord's work." Secular Jews are free of these demands. This situation does not apply at Keshet School. The timetable is defined according to the understanding that the religious side must pursue its traditions, while the secular side creates alternative traditions and content. In practice, through encounters with the religious side, the secular side creates what is almost a new "religion"—a religion that focuses on the class assembly, in which each student must take part, speak their mind and listen to the others, feel part of the collective, and seek spiritual content. This brand of secularism favors poetry, dialogue, intellect, and a rich emotional language. It could be termed a "high" brand of secularism. The participants are involved in a joint search for "meaning" as defined by Frankel and Maslow, and they develop tools for discourse, self-expression, and attentiveness.

The religious ceremony does not necessarily create a "high" experience. The cyclical ritual of the prayers does not demand a great intellectual effort, attentiveness, or originality, although intent—when it is present—is highly valued. The secular assembly, by contrast, has a creative character that allows change within a fixed structure and thus demands a greater emotional and intellectual effort. The school does not create room for a secularity based on exemptions. The religious students have to pray, and their secular peers have to attend the assembly. In real-world encounters between religious and secular Jews, the secular side feels exempt from the obligations incumbent on the religious. In Keshet School, the secular participants are committed to their own ceremonies, which may tentatively be regarded as an alternative religion.

Prayer, Assembly, and the Sense of Belonging

Gradually, the assembly also comes to embrace the religious students. When the class comes back together after the separate prayers and assembly, the teacher continues to discuss issues that are of concern to both sides. The school seems to give the students a sense of belonging that they find very important, and which is lacking in most educational frameworks. The morning ceremonies, followed by a half-hour discussion by the whole class, remind each student—frequently but informally—of his/her place as part of the community, thus adding a high level of cohesion to it. Children come to school with the feeling that if something is bothering them, they can share their problems with others. The school fights the tendency of the parents to seek to cut these hours in order to expand study time due to the "impending" matriculation examinations (from as early as the ninth grade).

Dubi, principal of the junior-high section

We run standard study classes, but we are also very concerned to maintain the morning assembly alongside the prayers. At the

> beginning of the school day, the whole class meets together for a few minutes, then divides up. After the assembly and prayers, the students come together again, and the lesson sometimes begins with an informal discussion relating to themes raised in the assembly or in the prayers. The basic assumption is that in modern life, when the students are exposed to pressure from political elements, the media and so on, they need much more time to process the challenges they face. A weekly "class period," usually devoted to organizational issues, is not an adequate framework for clarifying these problems.

The school gives its students something that is considered a rarity in modern times. Our identity is built in part through our functions or through the scenarios we experience each day as we attempt to give meaning to our lives. Adolescents are strongly influenced not only by their parents and their close friends, but also by the members of virtual communities based on the various channels of mass media. The ethos depicted by the media also has an influence on people who in turn influence the adolescent. This is an ethos of egoism, unbridled consumerism, and admiration of image over content. Children and families spend many hours in front of the screen, and real conversation becomes a rare commodity.

By devoting the first hour of the morning to an assembly, Keshet School is adopting an approach that opposes the culture of consumerism and ratings, restoring the communal and human dimension in the students' lives. The process grants weight to the students' own voices and enables them to examine in a profound and critical manner events witnessed via the media. This is an intellectual protest in an antiintellectual world, a spiritual protest in an antispiritual world, a protest of belonging in a world of alienation.

The reader may detect a hint of elitism in this value system, though the school works hard to avoid the pitfall of elitism. An effort is made to avoid creating a distinction between students from prosperous neighborhoods and those who come from less salubrious areas. The same approach explains why the school does not encourage tenth graders with poor achievements to move to senior-high sections elsewhere, as is often the case in Israel. It is important to recall that some 30 percent of the students at the school come from the less well-off neighborhoods.

> Dubi
>
> We do not usually ask students to leave at the end of the 9th grade, because we see the whole school as one unit, from pre-school

> through 12th grade. A child may be pleasant or unpleasant, stupid or smart, as the students put it, but we do not relate to these definitions. We are committed to social change, and thirty percent of the students come from poor social backgrounds. Despite that, we have an elitist image, which is a pity. We have strong integration in the 8th and 9th grades. Children come here with all kinds of problems, not necessarily ones relating to socioeconomic status, but Keshet School is a heterogeneous place and every child can finish their studies here with some form of matriculation certificate, full or partial. This is our approach.

The educational staff object to the growing emphasis on competitiveness as a part of the central ethos in an education system. There are more than a few "good" schools that have developed high educational pretensions and declared their support for the concept of a community school, while at the same time forcing half their students to leave at the end of ninth grade. There can be no greater gulf than this between pretension and educational reality. In most cases, this approach directs only a minority of students to tracks that ensure success and advancement, while the remainder are destined to fill boring positions in the bottom ranks of the job market.

The children at Keshet School generally express satisfaction with the spirit of the school, which they often compare favorably to "ordinary" schools.

> Diary
>
> During recess, I sit with A, a student in the 6th grade. Around us, a few boys from the junior-high section play ball. The children visibly come from diverse backgrounds. A tells me: "Before my friend came to Keshet, she studied at R [a school in a prestigious neighborhood], where they used to shout at her and give her a lot of homework. It's as if they don't care about the children in those schools. Here we have activity groups, sport and art. The teachers care more about the students than in other schools. There is violence, but not too much, and mostly in the junior-high classes. As you can hear, the students here curse . . .
>
> I try to ask A about the different kinds of students. She says: "There are students here from all kinds of places, but there isn't any difference between the children." A moves away, and I ask the same question of G, a physical education teacher who happened to be walking through the yard at the time, and who already knew me

> from another school. He commented: "There are children here from the Katamonim and the connection is really good. The connection between boys and girls at the school is also good—that's something I haven't found in any other school. Most of the children here are very good, although there are also others."

When a school abandons the ethos of excellence as its central goal and seeks to provide a warm, human community, a real possibility emerges of creating fruitful dialogue between children from very different backgrounds. "The encounter with the other"—a common mantra in the school—is not just a slogan. The school finds realistic ways to turn this mantra into a practical ideology. One of the main criticism of American consumer culture, which has become a dominant culture in Israel, is that it encourages addiction to oneself without regard for others—a narcissistic culture in which the happiest person is the one who "dies with the most toys." The feeling is that capitalist discourse does not relate to the need for individuals to contribute or belong to society in order to meet common needs (Etzioni, 1994). In the schools we have examined thus far—Neve Shalom, Bialik, and Keshet, where the morning assembly emphasizes themes of communality—an attempt is being made to put the individual's membership of the community back in the center of the educational experience.

The Question of Blurred Identities

Despite the effort to reinforce the students' diverse identities, while accepting and understanding the other, Keshet School nevertheless appears to create a more complex identity that internalizes behaviors that cannot be categorized simplistically as either "secular" or "religious." In order to prevent confusion, the school takes care to remind students, both religious and secular, of the each group belongs to. Students may visit the assembly or prayer service of the other group, but no more than once a week, reflecting the desire to avoid confusion and identity problems among the students.

> Dikla, former head of the elementary school
>
> The founders of the school assumed that if the student's personal identity is clear, they are secure and they can participate in the assembly. If the children live in a ghetto, they cannot be secure in this way. For example, I grew up in a secular ghetto (a kibbutz).

> People who live in this school get to know the other. One of the goals of the assembly is to reinforce the children within their own group of affiliation. In other words, if something dynamic happens in the prayer group, this should be mirrored in the secular assembly. It is important to give the child a sense of belonging. Secular and religious children can visit the assembly or prayer service of the other group, but they know where they belong, just like in a family. That is why they can only visit the other side once a week. If a child spent every day with a different family, they would become confused.

These comments are reminiscent of remarks we heard in Neve Shalom School. As already noted, however, the division is clearer in that school, since there is virtually no realistic chance of cultural conversion. Keshet School, however, despite the declarative division between the groups, leads to the emergence of complex and diverse identities. The school promotes the creation of a new Jewish culture that leads to a higher level of identification with Jewish tradition, rituals, and symbols among the less traditional and iconographic section of the participants. Conversely, this new culture leads the more traditional and iconographic population to open itself to "softer" manifestations of the formative Jewish myths. It is possible that Keshet School actually manifests a type of Judaism that already exists in Israel, albeit without a name—a traditional Judaism that cannot easily be defined according to the usual poles of orthodoxy and secularism. Although this Judaism lies somewhere in the middle, it also differs from the Conservative stream as seen in the United States. Ruti relates to Sheleg's book (2000) on the "new religious" in attempting to define this approach:

> In ideological terms, they range from moderate Orthodoxy to traditional Jews with a commitment to Halacha. It is something very powerful. There are also quite a few Orthodox people at the school who do not wear a skullcap, because of the very clear symbolic message this sends. I am not involved in the diffuse definition of traditions that include an element of ritual behavior. For all the diversity in the school, it is important to define borders.

The school seems to find it difficult to cope with this "middle-of-the-road Judaism." It is present, but it cannot easily be conceptualized and thus remains on the covert level. There may be a conscious or subconscious fear that defining the school as "traditional" would upset Orthodox parents.

Dubi

A complex dialogue takes place between what we want and what we have. In general, the religious public is very cautious. Religious parents want their children to have what they consider a sufficient profound religious education, but some Orthodox parents come here precisely to avoid closed education. We do not have a survey on this aspect. But it is interesting to see that a religious boy in the 9th grade, who does not engage seriously with his spiritual world and prefers to deal with trivia, makes friends with a secular boy who has a similar approach. By contrast, a religious boy who is seriously interested in spirituality finds common ground with a secular boy who is also preoccupied with his spiritual world. The connection can be vertical rather than horizontal. It's true that we have a problem defining what we mean by "traditional." We have some students who fall somewhere in the middle and defining it isn't easy. The students who are defined as religious study more Halacha, and the older grades devote more time to Jewish subjects. They cannot make do with the minimum. There is a problem when a boy whose parents are not Orthodox studies in the religious group. We are aware of the problem, discuss it and engage in dialogue.

Tova

There are all kinds of stories here about students who are considered religious. There are also mixed families, or parents who are either becoming religious or secular, as well as families that cannot be considered Orthodox in the full sense of the term. There are also couples becoming religious. The word Keshet ("rainbow") is appropriate given the diversity here. For example, I have students who go to the Reform synagogue Kol Haneshama, are committed to prayer, but after the service finishes on Shabbat they drive off in their car. So what should I do, tell them not to pray? That's why Keshet is a very appropriate name for the school.

The ideological approach of Keshet School may suggest the possibility of the emergence of a new type of Judaism that is complex, pluralistic, and critical, one that could develop from a network of schools offering a tolerant encounter between both sides. Many parents and students would seem to be ready for such a process. A similar phenomenon may be seen in the development of the Tali[6] schools—state schools with a positive approach to Jewish tradition. There are some similarities between

Keshet and the Tali schools, although Keshet puts greater emphasis on providing a religious experience for the participants.

Ruti

There is a similarity between Keshet School and the "Tali" schools, which devote more attention to Jewish content than regular state schools. Keshet School offers a more profound manifestation both of secular and religious identity. We accept families that travel on Yom Kippur and eat bread on Pessach, as well as families that strictly observe the Halacha in all areas of life. Tali does not relate to distinct cultures, whereas for us we are not just talking about a more Jewish-style atmosphere.

Our school has a proper representation of both the distinct identities. When we wanted to give all the 2nd grade children a prayerbook a few years ago as their first book, there was an uproar among secular parents who were fiercely opposed to this. That wouldn't happen at Tali. At the same time, we have to cope with a Halacha that teaches that eating bread at Pessach will lead to a Divinely-imposed death penalty. The religious side at Keshet is aware that there are children here who eat bread at Pessach, or who eat on Yom Kippur, or who go on vacation in Sinai on that day.

The Keshet option is rejected both by parents who fear that its openness to secularity will erode Orthodox values and by secular parents who view every manifestation of the Jewish religion as a symbol of coercion and backwardness. Keshet School provides a congenial atmosphere for both secular and religious Jews who not only have a distinct identity but also identify with the idea of the encounter as a tool for learning and growth. In addition, and paradoxically, Keshet School is also a convenient framework for Jews whose attitude to Judaism and Jewish tradition is ambivalent. In terms of behavior, these participants in the school tend toward orthodoxy; Ruti Lehavi refers to them as "the fourth stream of Judaism." For many of the students at Keshet School, internal contradictions do not seem to be regarded as such.

Diary

I visit a family that is regarded as Orthodox by the school and by the children's own definitions. The parents are willing to drive on Shabbat, but refrain from doing so because their children object. We sit outside and eat Shabbat lunch. Y, a religious girl from

> Keshet School, arrives unexpectedly on her bicycle. She says that her family does not drive on Shabbat, but answers the telephone. I do not sense any confusion or apology in her tone. When I ask her about her religious practices, she tells me that there are "lots of children like this," particularly her friends from Keshet and from other religious schools.

Keshet School is home to a plethora of complex Jewish identities, from secular cultural through moderate orthodoxy. The pluralism of identities found in the school is reminiscent of the situation outside Israel, where the liberal streams of Judaism—Reform, Conservative, Reconstructionist, and others—are much stronger than in Israel. The school may actually reveal the distortion behind the religious-secular dichotomy. Keshet School illustrates the possibility noted in the study of Levy and Katz (1993) that many Israelis are actually traditional, observing parts of the tradition as they wish and choosing without any sense of guilt. Keshet offers a framework for this type of Judaism. To return to Dubi's comments above, the encounter provided at Keshet is not so much between religious and secular Jews as it is between individuals with different religious styles and tendencies. In this encounter, those with a more spiritual inclination are attracted to each other, as are those with more practical or mundane orientations, just as sports lovers congregate together. The imposed distinction between religious and secular is not observed in sociological terms.

The school works hard to ensure that all the bands of the rainbow feel at ease; a religious student should not feel uncomfortable and a secular student should not feel threatened by religious coercion. The school prefers to accept students whose families define their identity as either "secular" or "religious," and whose value systems do not negate the other identity.

Ruti emphasizes that the school is not interested in "a comfortable and pleasant middle-of-the-road position; we do not want to be Tali." From her perspective, the litmus test is to ask where the parents would have sent their children if Keshet had not existed. She claims that families would do better to send their children to a regular state school, or to a state-religious school, rather than to Tali.

The Curriculum

The school has been obliged to show a large measure of creativity and originality in developing its curricula. Israel has curricula for the state system and for the state-religious system, but there are no programs combining the two. Keshet School is developing an alternative educational program that will meet the needs of the unique encounter created in the school.

> Ruti Lehavi
>
> For the past two years, we have been in the process of developing unique curricula for Keshet on the subject of the High Holydays. The curricula were developed jointly by the teachers and representatives of the school association, with the assistance of an instructor from the school. In the senior-high section, we use three special curricula, in the fields of the Oral Law, Bible and Prayer. We felt the need to examine the provision of pre-service or in-service training for our teachers. We are examining different options in this respect. We also thought about establishing a teacher training program in the spirit of Keshet. We have not managed to do this yet, but we are still looking into the idea. Our new curricula relate to different perceptions of Judaism, not just Orthodox and secular, but also the Reform and Conservative streams. Exposure to different streams of Judaism is important, and the curricula will also be useful for other schools.
>
> Even when religious and secular students study separately, the children are exposed to different cultural approaches. Religious schools emphasize the religious contents and secular schools the opposite, but the students can still learn about the other approaches.

Thus we see that the curricula also relate to parallel value systems as valid options for human behavior. The values are clarified and a wide range of possibilities are offered, beyond those presented to students in the regular state schools.

Some parents complain about particular aspects of the school's work.

> A mother from a liberal, educated Orthodox family
>
> The encounter between religious and secular children is very important. I have no doubt that there are profound issues to discuss and there can be real content, beyond the mere fact of the encounter. It would be worthwhile developing more content—they don't have curricula yet. There is a lot of improvisation in the studies. There are wonderful people here, but things move very slowly. We shouldn't have to rely on curricula from the state and

state-religious systems—we could definitely create our own curriculum. The encounter itself is excellent. It expands the borders of identity; even if an element of uncertainty is introduced, that doesn't bother me. At some point they will have to teach more focused content, even in the academic subjects.

Many Israeli parents who belong to the "fourth stream" send their children to state-religious elementary schools so that they can absorb a little tradition and "Yiddishkeit." When the time comes to move on to senior-high school, however, they transfer their children to the state system. In the quote above, the mother has no problems with the encounter in the elementary school, but she would like to see its religious atmosphere and content stregthened. As a result, she intends to transfer her child to a state-religious school when he begins junior-high school.

Dubi

In the junior-high section, we aren't reinventing the wheel in terms of the content of studies, but we do change the emphases. We present the options of Hellenism and Judaism in the Second Temple Period as equally valid—what did we accept, what did we reject, what did we fight for, and so on. In the 8th grade, the subject is Judaism and Islam and their different influences. We also emphasize what we gained from it, such as the whole question of Maimonides and Aristotelian philosophy, which came to us via the Muslim philosophers. In the 9th grade, the subject is Christianity, including Christian hatred of Judaism and the absorption of Christian ideas—such as the existence of heaven—in Judaism. In the 10th grade, we move into a more standard track of Ministry of Education curricula. However, as an experimental school we enjoy the privilege of developing some special curricula.

These comments explain why the content and style of learning in the school may deter parents with iconographic and particularistic approaches. The myths "created" at Keshet are "soft" myths that do not negate the other or promote ethnocentrism, but they rather promote positive dialogue with other cultures. The idea that Judaism has learned and adopted elements from other religions and cultures is revolutionary for many Orthodox Jews. Critical examination of your own culture by comparison to another culture encourages a more relativistic approach to both, and this relativism allows you to take what is best from your own culture and from the other culture as well.

Keshet School offers interesting models not only in terms of the content of studies, but also in terms of style.

> Diary
>
> I arrive at the school at 9 o'clock in the morning. In the staffroom, some teachers are discussing the news that a number of well-off families have decided to buy or rent homes in the neighborhood so that they can be closer to the school. Y smiles at me and asks if I would like to observe a Beit Midrash class. I have heard a lot about the Beit Midrash lessons in the school, and follow her. A girl in a belly top passes by, and I hear her comment to her friend, "It's OK, we don't have to rush—it's only Beit Midrash now." A student tells me: "Here we learn about people, homes, streets, about Israel and the world, and then we divide into pairs and use drama to tell the other what we learnt. We're study pairs learn together." The teacher asks to say the Blessing after Meals, because they just ate. A religious girl tells me that "sometimes the secular children say more blessings than we do." I listen to the singing; the tunes remind me of a Jewish summer camp I attended in the United States.

The teacher tries to remind the students what they learned in the previous lesson. I admire her willingness to invite me into a class without prior arrangement. She talks to the students about the meaning of the concept of "home" and its connection to the fields of culture and religion. "Today we are going to continue to discuss the idea of home, but in different ways. Each of you will choose how to express your own understanding of what home means, but within a study group. At eleven o'clock we will divide into workshops." The children show an impressive level of self-discipline. The study topic reminds me of the research projects I saw at Bialik School, where the children also discuss concepts such as home, time, and borders. They divide into groups—two groups in one room. One child talks to his friend in English. Four girls study together and tell me that they learn by studying a particular text. I ask why some of the students are in small groups and some in large groups. They explain that they started as part of a larger group, but then they broke away. I point out that one boy is on his own. "He prefers it that way just now," one of the girls explains. "That's how he learns. Sometimes I also sit by myself." They tell me that they "like this class because there is just text, and no teacher, and we learn together." "We have to manage by ourselves and choose how to express ourselves." I go outside and see two girls sitting together. One is religious and the other secular, although there is no difference between them in terms of appearance. They are preparing a dramatic presentation based on the text.

Another student tells me that he comes from H, a moshav. He prefers to study by himself. "Sometimes I like to work with people, it all depends on my mood. There are all kinds of different workshops and channels for creativity and ways to express ourselves."

As in Bialik School, we see the use of cybernetic learning that emphasizes the acquisition of knowledge among the participants by means of value-based negotiation reflecting different perceptions of a single reality. This approach shows a high level of consideration for the children's individual personalities and learning styles and enables them to present their understanding of the subject in diverse ways. It is interesting to note that the traditional Jewish "Hevruta" (study pair) blends well with modern perceptions of "cybernetic knowledge." Torah study is the ultimate act of interpretation: knowledge changes from generation to generation, and each generation brings its own unique perspective to the text and to the body of interpretation. Keshet School trains the students in independent learning, research, and diverse forms of expression.

When some of these students enter higher education in the future, it will be interesting to see whether they still favor critical and relativistic approaches toward cultures in general, and toward Judaism in particular, and whether they will remain tolerant and empathetic to others. How will they analyze the canonical Jewish texts and other works? It may be assumed that the open and bold approach they have acquired to interpretation will enable some of them to bring profound and complex dimensions to the texts, helping to make them relevant for socialization in modern Jewish culture.

Attitudes to the Other

The school's often-repeated mantra of the encounter with the other is reflected in the behavior and attitudes of the students. Although Keshet School offers an encounter within the Jewish context and addresses questions of identity, a conscious effort is made to expand the encounter to additional populations. The school runs encounters with Arab schools, with varying levels of success, although organizational problems are encountered. The students support these encounters.

The school also enables the children to meet the other through encounters with a school for children with disabilities. The children work quite hard in this field, showing an open-minded and serious approach. They draw on the language of the school in describing these encounters: "We met people who look different from us on the outside, but essentially they are very similar to us."[7] The encounter with the other is one of the foundations for implementing the personal commitment program in which the students at the school participate. It encourages the teachers to find additional channels for this encounter, and to extend it beyond the confines of the religious-secular encounter between "regular" children.

The School and the Community

The school is a relatively new presence in the community and is taking the first steps toward integration into communal activities. Though it has not yet reached the point where it can exert an influence, it is no longer seen as an alien or elitist presence by the residents of the Katamonim.

> Dubi
>
> The connection with the community in which the school was established is still at an early stage. At first we were strangers. Today, there are a large number of students from the neighborhood in the school—one-third of the students come from the neighborhood. It has become a legitimate and popular option. We want to improve the appearance of the street and parks, hold an open day—we have all kinds of ideas. Children from the neighborhood come here in the afternoons. We try not to remain aloof. The school is still not closely enough involved in the life of the community, but we are making serious efforts toward integration.

Thus the idea of the encounter is not confined to a static situation within the school. An effort is being made to extend the ties with the community and to avoid being perceived as an alien presence.

When it first opened, Keshet School was situated in the Givat Ram neighborhood of Jerusalem, which Ruti Lehavi feels was ideal for a school drawing students from throughout the city. Givat Ram, which is situated close to the city's museums, the university, the Knesset, and the Supreme Court could all convey a message of neutrality, since "the area doesn't really belong to anyone, or if you prefer—it belongs to everyone."

However, the staff seems satisfied with the current location. The hope is that a large campus will be built, housing a cultural and leisure center and providing space for meetings and study programs. Naturally, this program also includes the students, teachers, and parents in the school, regardless of where they live. According to the plans, the campus will include a library, sports center, Beit Midrash, teacher training institute, and so on.

The school works hard to involve the students' families in its daily life. Various frameworks have been established that involve parents in joint studies. Parents are also involved in the education committee and in writing curricula. Some parents have even taught at the school on a voluntary basis. One parent, a psychologist by profession, gave the teachers a lecture as part of a workshop on the subject of "identity".

Plans for involving the parents in the life of the school have run into problems due to geographical distances and the fact that many parents are very busy. Despite this, the staff continues to make efforts to involve them whenever possible.

The Teachers in the School

The teachers in Keshet School adopt an open and pluralistic approach to education. They learn to see themselves as active mediators in the educational process—as partners, rather than formal transmitters of knowledge. As we have seen, the teachers enable the children to study independently, recognizing that different individuals will use different learning styles and find their own ways to acquire knowledge. It may be that teachers who share this approach are tacitly channeled to the school. Rumors about the special nature of the school spread by word of mouth, and the school uses its connections with the David Yellin College of Education in Jerusalem to recruit teachers who can integrate in the school and cope with the special challenges it presents.

As teachers gain experience in the school, they internalize the school's approach to work and its unique ideology. They are expected to be able to lead the encounter with the other, since they function as key agents for socialization within the school system.

Not all the teachers find it easy to cope with the study methods and approaches used in Keshet School, particularly during the early stages of the work. Some of the religious teachers seem to find it difficult to cope with the blurring of identities (or their enhanced complexity).

Diary

I stand in the small coffee corner and make myself a cup of coffee. I chat with a teacher who was trained at a religious seminary. "The diversity here is interesting, but the school is slightly chaotic. Maybe this is suitable for a certain population. The affiliation of the students is chaotic—it isn't always clear who is religious and who is secular . . . Everything is open here, including the style of learning. I think we need to set clear limits. A lot of effort is invested in the children and programs, but I think that the school kind of confuses the children. They need a clear line, and I'm not sure that's what they get."

A Religious Teacher

The encounter with the complex identities of the students in Keshet School is not a simple one for the teachers. It raises questions of identity that many of us also grapple with. The complex preparations of the studies and the need to invent innovative curricula are reflected in the internal discourse of the teachers, as is the constant attention to issues relating to the encounter with the other.

The combination is very special. All my life I taught in religious schools. Here it is half and half, and there are also traditional students. Personally, I am also in a period of examining my own identity . . .

I live in a mixed religious-secular community where there is also a school that is attended by both religious and secular students. But the definitions are much clearer there. It is obvious to the religious students that they learn more Gemara and more Judaism. I didn't even have a chance to get to know secular Jews until I went to the army.

Management of the School

As mentioned, the school began at the initiative of Ruti Lehavi. She sought to establish a framework that would reflect her own belief in encounter as a tool for learning about oneself and about the other in the Israeli-Jewish context. Her search for the ultimate model for this encounter may be connected with the fact that she herself, as a religious Jew, lives harmoniously in a mixed religious-secular family, although she claims that this hypothesis is irrelevant.

Ruti, who is now returning to the school to run the senior-high section, and Dubi, principal of the junior-high section, taught together for many years at the Experimental School in Jerusalem, and even ran the school. Many of the management and study techniques they have brought to Keshet were tried and tested in the Experimental School, which has undergone several changes in its direction as it constantly explored the boundaries of experimentation and openness. There has always been an element of tension in the Experimental School between radical educational approaches and the need for at least a measure of integration into an institutionalized education system that has not always taken a favorable view of educational experiments.[8]

An important Experimental School principle that has been adopted in Keshet is to maximize the attention given to each student, regardless of his/her achievements, status or popularity. In his book, Zimran explains this principle:

> One of the teachers at the Experimental School relates: "What impressed me was the fact that a large group of teachers spent over thirty minutes discussing a student who was giving them a rough time—not in an effort to expel or control him, but with the hope of finding a way to draw him closer and develop a recovery program to be implemented by the very teachers he was aggravating." He adds: "The group discussed his problems with a sense of respect and out of responsibility for his future. The same was true when the discussion turned to other students."[9]

Echoes of this approach may be found in Keshet School. In addition to the efforts made by teachers to get to know the students and develop meaningful relationships with them, students are also involved in making decisions on significant issues relating to the school. For example, a group of students presented their needs to the staff as part of the planning for the new senior-high section in the school.

> Dubi
>
> We established small discussion groups with the students, for example five students from the 9th grade, and then we discussed the issues with the parents. We held a lot of discussions and we will try to implement the points that were raised. We are planning a high individual study program for the 11th and 12th grades which the students will build themselves. We take the students into account on many aspects—that is the advantage of a small school, apart from the obvious drawbacks.

Keshet School has adopted another feature of the Experimental School, namely heterogeneity as an ideal and a value rather than a constraint. Ruti, Dubi, and other senior figures in the school do their best to inculcate this ideology, despite the difficulties it creates, since many teachers find it easier to work in homogenous classes. Keshet School deserves the "rainbow" epithet as a reflection of not just its outlook on religion, but broader aspects of student diversity as well.

Conclusion

> Dubi
>
> We encounter some very paradoxical situations here. In the 8th grade, for example, one boy who grew up religious and attended religious schools realized that he did not want to observe Shabbat, and then learned that his brother works on Shabbat. In the same class, one of the girls who is defined as secular observes Shabbat, eats Kosher—the whole package. Israeli society is seen here with all its complexities. Our definition is simple. We believe that everyone needs an affinity group, and the school must declare that it is developing religious and secular individuals. If we start to mix and match, we will be more like the "traditional" school (the elementary, junior-high and senior-high schools of the Tali network), or other schools that place an emphasis on Jewish studies. We often discuss issues relating to real-life problems. Last year, a student returned with his parents, who had worked as emissaries abroad. He celebrated his Bar Mitzvah in Moshav Shoresh, outside Jerusalem. He invited everyone, but the religious students (who could not travel to the event on Shabbat) told him "How can you do that if you're our friend?! How can you have your Bar Mitzvah on Shoresh when we can't get there." The parents had not taken this into account. He was almost ostracized, but neither he nor his father could understand why it was a problem.

Keshet School offers a unique educational opportunity in the Israeli educational world. The school's ideology of Repair seeks to narrow the gap between religious and secular students by bringing them together in a single educational framework. The assumption is that a positive encounter will lead to familiarization, study of each other's sources, and thus mutual acceptance, while also strengthening the distinct identities of the participants in the encounter. The school meets a real need of many parents and students for a good school that views diverse ways of life as an advantage rather than a problem.

The school enjoys a higher level of understanding and acceptance than is usually found among secular and religious Israelis alike. The interesting profile of the student population, the fact that many parents are themselves examining their own identity, and the presence of "mixed" families in the school, with all the complexities this raises in Israeli society, enable the children to develop complex identities that go beyond the standard definition of religious or secular. The relativistic and critical approaches to textual study help the students navigate these complexities and enrich their Jewish perspectives. The school grants a large measure of autonomy to the families in finding their own self-definition. There are also students who come from traditional or Reform families that observe the Halacha partially. Despite this, the school addresses both ends of the spectrum and prefers a clear definition, in order to prevent a situation where "full-fledged" religious or secular children feel uncomfortable.

The encounter in Keshet School is an egalitarian one permitting mutual learning and the softening of stereotypes. This meets the conditions proposed by Amir[10] for an effective cross cultural encounter. It should be noted, however, that the differences between the participating populations are much smaller than those in Neve Shalom or Bialik School. Conversion of identity is not perceived as a traumatic or tragic development. In practice, the school offers an entire range of complex identities that may be considered and chosen by the students. One would expected the question of development of identity to be critical among the adolescents, since children of this age are intensely involved in issues of identity; this is also an age when increasing importance is attached to the peer group in the context of the development of identity.[11] It is possible that Keshet School develops a clear identity; it is equally possible that it helps represent a complex traditional orientation that is present in Israel, but one that is largely unrepresented due to the dichotic and partially artificial divide between the religious and the secular. The symbolic dichotomy may be much stronger than the actual situation, as manifested in everyday encounters between religious and secular Jews in Israel; however, these weaker voices are not usually represented in public discourse, which takes place primarily in the media.

Studying in Keshet School rather than other schools offers students a chance to examine Jewish culture by choice. The complex examination of Jewish culture, addressing humanistic, critical, and religious approaches, and the emphasis on the dialogue between these approaches help promote the choice of a more complex identity, removed from the extremes of the spectrum. Despite this, Jewish identity remains very strong. This complexity may offers advantages in an increasingly complex world that demands of individuals a high level of reflection, adaptation to change, and acceptance of the other.

The staff in the school take the "encounter" label seriously and continue to search for additional partners in this process—Arabs, children with difficulties, and the community—although much remains to be done in these fields.

The teachers and principals see themselves as active partners in maintaining the school as a framework for encounter with the other. They attempt to provide a sense of belonging, support, and acceptance for all students, regardless of their academic achievements. The school generally adopts an informal and open-minded approach.

Judaism by Choice

> A Jew is a person who has one Jewish parent and who identifies as Jewish, accepting the responsibility and the privilege of being a Jew.[12]

Many Israelis are defined as Jews in objective terms. In a broad constitutional sense, Israel is defined as the state of "those who are unquestioningly accepted as Jewish by some Jews, even if not by all."[13] The Law of Return is disinclined to enter into ideological examinations, except in cases wherein an individual has explicitly converted to another religion. The law includes in the Jewish people individuals who are not considered Jewish from the standpoint of the Orthodox Halacha (such as Reform converts and the grandchildren of Jews). This definition, which is developed in constitutional terms, is broader than the narrow Orthodox definition. Keshet School is an agent of socialization for Jews who choose voluntarily to address Judaism in the religious, cultural, secular, or complex context. Choice is a key theme in the school. Students are presented with educational texts, and the teachers—as active agents of mediation between the text and the learner in the fields of knowledge and interpretation—enable the students to choose their identity through a process of genuine research, introspection, and free choice.

With its unique Ideology of Repair, Keshet School may provide a positive model for a society that could emerge in Israel as and when more positive conditions are created for the encounter between religious and secular Jews. It could serve as an example for a future educational approach in an era of peace and after the separation of religion and state. It may be assumed that Israelis, once freed of the fundamental problems of survival, will be more inclined to devote serious thought to the complex task of creating new forms of Jewish identity.

CHAPTER FOUR

Kedma School

(7th–12th grades, approximately 170 students)

Much has been written and said about the difficulties faced by the Jewish immigrants who came to Israel from the Arab countries. These immigrants found it hard to integrate into Israeli society and secure positions of power, and it is now widely accepted that serious mistakes were made in the process of absorbing these immigrants. In the educational field, the gaps are particularly wide. At first, little attempt was made to integrate them, and the immigrants remained in their communities with a poor education system. In the 1970s, efforts were made to promote integration in schools, based on the assumption that the encounter would automatically enrich the weaker students. Some students indeed benefited from integration and enjoyed mobility within Israeli society, but many were left behind. In a large number of schools, streaming policies concentrated weaker students in a single class, and in most cases these students were Mizrachi Jews from disadvantaged neighborhoods.[1] In many areas, integration seemed to be a failure in social terms. At the end of the 1990s, a number of incidents highlighted the tensions between students and young people from different socioeconomic backgrounds. In one instance, students from well-off families attending a prestigious school demanded that the buses they took to travel to and from school be separate from those that their poorer classmates took. On several occasions, the media discussed cases of bouncers at discotheques refusing to allow entry to young people whose physical appearance and/or dress identified them as "Mizrachi."

Integration was a failure for many reasons, most notably because it did not create an *equal* encounter. One hegemonic culture was perceived as desirable and worthy of imitation, while another was stigmatized as vulgar and best forgotten. The curriculum taught Jewish history from an Ashkenazi and Eurocentric perspective, almost completely ignoring the rich history and experiences of Jewish culture in the Arab countries.

The favoring of parents from strong socioeconomic backgrounds who belong to hegemonic ethnic groups is found in schools around the world. Studies in the United States have shown that the lighter the color of a student's skin, the more favorably they are treated by teachers (even if the teacher is dark-skinned).[2] Discrimination exists on many levels. Hammou[3] examined the attitude of youth interrogators toward young people questioned by the police and found a preferential and lenient attitude to young Ashkenazis ("Don't do anything like that again or I can't tell you what will happen") as compared to the attitude to Mizrachi youth ("You're a born criminal, all you guys are going to do time eventually").

The teachers and founders of Kedma School are well aware of these stigmas. In response, the school offers a corrective program of resocialization for youngsters who have been ignored and alienated by the establishment, and hence internalized the message that they were worth less than others. In sociological terms, this model can be justified in the context of an aspiring capitalist society, since through the process of socialization, whereby each person learns that their function in society is vital and worthy, some children are channeled to a future of success, management, and respect, while others are sent to serve the former in low-prestige positions devoid of interest, challenge, or an opportunity for advancement.

How does Resocialization Work?

Kedma School is situated in the Katamonim neighborhood of Jerusalem (as is Keshet). However, while Keshet School occupies a site in a relatively prestigious part of the neighborhood, Kedma is situated in the heart of the local housing projects.

Diary

The neighborhood is fairly run-down, despite the efforts of Project Renewal, an urban renovation program introduced by the Likud-led government in the 1980s. Posters and signs in the neighborhood reflect support for the political and religious leaders of the Mizrachi community in general, and Shas in particular. Sometimes a child wearing the colors of Betar Jerusalem soccer team (whose support comes mainly from the Mizrachi working-class population of Jerusalem) walks alongside a skullcap-wearing adult. Occasionally I notice migrant laborers in the street; as in Tel Aviv, their presence is evidence of the socioeconomic status of the area.

The Jerusalem Education Authority has begged Kedma School to relocate to another neighborhood, but the leadership of the school is convinced that it should operate in the neighborhood where most of its students live. Like Bialik School, Kedma also attaches great importance to its bond with the local community.

The resocialization at the school, or the ideology of Repair it seeks to implement, argues that, as the result of ethnic discrimination and socio-economic status, many students do not enjoy equal opportunity within the education system. The school aims to correct this situation. The school also argues that the efforts in the education system to promote "integration" have not given these children an equal opportunity or respected their culture.[4] The children who come to the school have poor self-esteem in two areas. In terms of academic achievements, they feel that they are poor students, because their achievements to date show that they are not as good as others. In terms of internalizing oppression, they have a vague sense that something about themselves and their culture is perceived to be wrong. The internalization of oppression is a characteristic of a society controlled by relations between the dominator and the dominated, as we also saw in the case of Neve Shalom.[5] As noted, the school staff see a sharp difference between the attitude of the establishment toward Arabs and its attitude toward Jews who immigrated from the Arab countries. It should be noted, however, that the school offers an unusually broad version of Mizrachi identity. There are students in the school who are not of Mizrachi origin (including students from Ethiopian and Russian families). Mizrachi identity becomes a catch-all for all those who have been oppressed or find themselves outside the mainstream of Israeli society. The discourse in the school emphasizes an analysis based on class and inequality.

Identification

Many of the Mizrachi teachers in the school have experienced inequality and discrimination, and have reached their current status despite these experiences. In many cases, their decision to work in Kedma School is based on a sense of mission. They seek to engage in Repair, and their work is rooted in a sense of identification. For their part, the students encounter in their teachers role models who began at the same starting point as theirs and have achieved success. This similarity facilitates the complex task of the teachers. The teachers monitor the academic success of each student throughout the year; much of the contact between teachers and students takes place through one-to-one conversations.

As in the other schools we visited, here too a foundation of trust is established between student and teacher. As in Bialik School, social and economic problems mean that discussions often extend to areas of general advice and life skills, and help is provided accordingly.

Through their encounter with the frustrations and difficulties faced by their students, the teachers reexamine their own school experiences, which were usually predominantly negative and dominated by a sense of alienation and failure.

> Teacher
>
> I remember when I was a teenager . . . When we went to the junior-high school we were all in shock. The other students were doing equations with two unknowns, and we were miles behind . . . We were in a race against time, and I had to catch up with all the material they had studied. It was hard for the guys from the neighborhood. We were together through the 9th grade, then we split up. From the 10th grade, I studied in the academic track in Denmark School. I didn't feel I was accepted. Some guys didn't make it, and whenever we met we used to ask what each one was doing, and I could feel their embarrassment . . .

These personal narratives motivate the teachers, reminding them of situations that they seek to avoid in Kedma School at all costs. The school makes an honest effort to ensure that the students, who are accepted without entrance conditions, are successful in their studies. The teachers work hard to nurture academic ability, a positive self-image, and cultural pride among the students. The high level of responsibility that the teachers accept may sometimes lead to burnout.

Coping with Burnout

The teachers in the school face a difficult and demanding task. Many boys and girls in the school show the anger they feel even today due to the poor treatment they have received from different quarters over the years.[6] Many of the students have academic, social, and economic problems that demand a broad-based support system. Over the years, the teachers in the school have been unsure whether they should involve counselors or psychologists in their educational work. With some justification, they feel that psychologists and counselors are sometimes used by the establishment as an instrument for screening and directing students such as those who attend Kedma toward blue-collar jobs. They also feel that these professionals tend to favor young people who come from a background similar to that of the system imposing the screening process.

Clara Yonah, the founder and principal of Kedma School, expected that the supervising teachers would fill the position taken in other schools by "therapeutic" professionals. Ariella,[7] a key figure in the school responsible for staff training, makes these comments:

> Clara, who has the energy of a bulldozer, and the founding team created practical policies as an ideology. One of these policies is that no-one is left out and we never give up hope. The result is that no student has ever been expelled and there is no red line that leads to expulsion. This principle still applies, but even so the limits are tested.

The school ultimately found ways to integrate therapeutic and ancillary professionals who were thought to have stigmatic approaches or cultural biases. The pooling of resources that is habitually created at Kedma is reminiscent of the efforts made by the teachers at Bialik School to help their students. In both schools, the goal is to help the student, and therapeutic components are intended to help students realize their potential. Both schools make an enormous effort to motivate the students to achieve progress, based on an almost unconditional belief in the ability of each student to learn if they are provided with the proper conditions.

Many teachers who work in Kedma School have questions about their own future. Some teachers originally intended to work in the school for a limited period, perhaps until the first class graduated from the school. Their work gives them a strong sense of satisfaction, but the level of commitment it requires leads to burnout. Positive feedback encourages the teachers to stay, but low pay is a disincentive for many, particularly male teachers, who feel that their status has been damaged and look for alternative possibilities. The students themselves also tend to evaluate the worth of the teachers based on their economic status.

> Teacher (male)
>
> I come to school in my old car and they pull out pricelists and laugh about how little it is worth. I tell them that I really don't care, but they don't believe me. They ask me what I am doing here. They spend hours fantasizing about what cars they want to buy. We were the same when we were seventeen. Materialism is hard to fight. They internalize the dominant values in society. But with all my love for this place, when I look at myself I wonder whether I am doing the right thing to work here.

Although the teachers are sometimes frustrated by the tendency of the students to accept Western consumer ethos uncritically, the students' comments show that they are influenced by their teachers' ideas as well. Rafi, the philosophy teacher, has a Bohemian appearance and is popular with the students. He often talks to them, "like the principal Clara," about the need for choice. For their part, the students internalize Clara's message of choice, using the example of Rafi's economic condition. "Well, that's his choice, give Rafi a room full of philosophy books and he'll sit there happily, just like someone else is happy to choose money and new cars." Thus the students are inculcated with values held by their teachers, even if this cannot always be seen immediately.

Management

Kedma School belongs to a network of schools (of the same name) that share a similar basic ideology. However, significant differences were seen between the different schools in terms of their strategy for achieving the ideology of Repair. Clara believed that the Jerusalem school, with the founding and leading staff, was the ideal place for her to realize her own educational vision. Clara has a strong personality, is charismatic and authoritative, and employs an open and completely informal style of discourse. Many teachers have sought the attention and guidance of the principal, who is perceived as the "guardian" of the school's ethos. In recent years, however, an effort has been made to involve the entire leading team in sharing the burden of management. This team includes a large number of teachers who are capable managers and leaders, and who can provide a reserve of future leaders for the network of Kedma schools that the ideological leadership hopes to establish, based on the unique model of Kedma in Jerusalem.

An effort has also been made to involve parents and students, particularly the first class of graduates, as well as members of the community at large, in the decision-making and policy-setting process of the school. As noted, the teachers and the management team seek to serve as a vanguard for the establishment of additional schools that would enable students around Israel to complete their matriculation certificates in a framework that respects their culture and background.

As in the case of Bialik School, the staff of Kedma School is also opposed to the current trend toward the independent management of schools. The staff views this policy as unacceptable, since it turns the school principals into fundraisers and measures their success according to their achievements in this field. Inevitably, schools in more prosperous neighborhoods will be more successful in this task and will secure more resources, thus widening social gaps still further. The staff in Kedma believes that every student is entitled to receive services and opportunities from the state on an equal basis. The teachers who work in the school feel that the main problem of education in Israel is not any particular management style, but rather the gap in the level of education provided for poor children as compared to their affluent peers. They also note the correlation between national and ethnic origin and the condition of students and argue that this gap can be narrowed only when all children in the State of Israel enjoy equal educational opportunities. The school works in the spirit of this belief.

The Critical Approach versus the "Proof of Matriculation"

In ideological terms, Kedma School operates in difficult and paradoxical conditions. It emphasizes Mizrachi culture and heritage, which are barely present in the ordinary education system, and criticizes mechanisms of streaming and discrimination, such as psychometric examinations, educational counselors, psychologists, and matriculation examinations. At the same time, the teachers are expected to enable their students to succeed in terms of the parameters set by the dominant group, an expectation that ensures its continued hegemony, even if these parameters are considered illegitimate. Despite protests from some teachers who felt that the students should be able to "decide for themselves" whether the matriculation examinations were important or not, the school has decided in favor of the approach that encourages students to pass their matriculation examinations as a means, albeit problematic, for success.[8]

Many parents also advocate the introduction of studies for matriculation, in the aspiration that their children will enjoy a better chance of advancement than they did.

> Diary
>
> I sit in the home of G with twelve parents and a teacher. We discuss various issues relating to the school and its ties with the community. Suddenly, the conversation turns to the question of the importance of university studies. The parents tell their own "horror stories" from their school years, and emphasize that they want their children to go to university and to enjoy better prospects for professional advancement than they themselves have had. The teacher comments that "we should let the children decide when it comes to the question of matriculation examinations and academic studies." The room falls silent briefly. Then G, the host and an active volunteer in the school who is a charismatic and impressive leader, responds to the teacher:
>
> *G*: Today a degree is really important as an entrance ticket to life.
> *S*: (the teacher): I have a degree, and many people who don't make more money than I do.
> *G*: That doesn't matter, it gives you the choice.
> *S*: But I didn't study education.
> *G*: Even if my son studies ancient history from way back when, he can still work as a teacher after training. So he'll have a real choice. Would you accept me as a teacher in Kedma? You don't earn a big salary, but you are doing what you like. I make more money, but I'm doing something I don't enjoy. If my son wants to be a construction worker that's OK, but only if he does it out of choice—first he needs to get a degree. A degree is an entrance ticket to life.

These comments reflect the position of many parents. The parents support the work of the school, largely due to their impression that it is affording their children the equal opportunities that they themselves were denied in elementary school. The teachers at Kedma tend to emphasize Mizrachi culture and the pedagogics of liberation. When they attempt to draw the parents and students in this direction, however, they encounter some problems, since most of those involved do not readily accept such a revolutionary message. Their approach seems to be influenced by a broad acceptance of the criteria for success as defined by the hegemonic group. The school reconciles itself to this "surrender" and attempts to realize its plans in difficult conditions, without abandoning its revolutionary message and the development of a more balanced curriculum that reflects the full richness of Mizrachi culture.

Two types of teachers may be observed in Kedma:

a) Those with a critical cultural approach, who attempt to draw their students into a liberation dialogue and to enable them to take a critical view of society as a whole. These teachers work from a position of identification and are well aware with narratives of pain from their own personal experience.
b) Teachers whose work is based on a more modest ideology. From a position of empathy rather than identification, these teachers attempt to give tools for success to students who have not been given an equal opportunity. These teachers accept existing social order but seek to enable their students to succeed within the society from an improved starting point.

Each approach exists alongside the other, since both make positive contributions to the children and help them to succeed. The students themselves, and their parents, adhere to the position of reconciliation to the existing order and are pleased that at least there exists a school that provides its students with an equal opportunity.

The Curriculum

The curriculum attempts to create a paradigmatic inversion. As victims of a certain form of cultural oppression, many teachers and students at the school are obliged to cope with a feeling that there is something "wrong" with their culture.

S, a teacher in the school, recalls his adolescence

> Take the example of the collection of music I have here. These are the same things that my father used to listen to. But at school, we listened to different music. We were too ashamed to talk about our own music.
>
> We would talk about rock and funky music. Some guys went the whole way with their love of Mizrachi music, emphasizing their rough subculture and identity. These were the guys who had long since abandoned any dream of integrating in the Ashkenazi world.

They were in the "dummies" class—the remedial class, 7–1 and then 8–1. They didn't have anything left to fight about, and they had no identity problem.

My friends introduced me to Led Zeppelin. I couldn't stand to hear the band—I had never listened to that kind of music. When my friends came over, I would take out the mellow Arik Einstein music that my father listened to, and rock music. It was ok to listen to the things my father listened to. But I hid away the Moroccan music.

The language and culture curriculum developed by the teacher in Kedma School over the past few years seeks to make a contribution to addressing the need to develop a positive identity within a culturally hostile environment.

Clara

We don't follow the classic division into literature, language and expression. We have developed our own draft reader—we chose the texts and the children like them a lot. We divided each class into three groups. The children learn in small groups of eight students. The students really enjoy these classes.

This is the same learning model we encounter in Neve Shalom, Bialik, and Keshet—a model of interdisciplinary and cybernetic learning that enables the student to search for his/her own personal voice through a dynamic encounter with the text. The text plays a mediating role, creating an experience of joint learning with peers and with a guide. Some of the texts are universal, while others address issues relating to Mizrachi and Arab culture.

Diary

I observe one of the "Language and Culture" lessons. The group includes ten students and two teachers who run the class together. The teachers have difficulty beginning the class, since one of the boys keeps interrupting. Eventually one of the teachers bursts out, "Shut your mouth already!"

The boy falls silent, and the teachers read a folk legend from an anthology by Pinchas Sadeh. Some of the children listen attentively. One boy comments sarcastically, "What a moving story!" The teacher says, "Let's start to learn the elements." "According to the fifth element," one student suggests, and his friend laughs. The teachers explain that the legend does not have time or space, and the character is not defined. They try to guide the children toward a position on the legend—is it positive, negative or neutral? The students find it difficult to understand these concepts. After some discussion, however, they get the idea and discuss the moral of the legend. One of the teachers comments, "This one isn't like Cinderella, it isn't black and white. This legend is a bit closer to real life." Then she begins to discuss the repetitions in the story. "Typological," declares one of the children.

Another boy declares, "There's nothing so hard about writing a legend."

The teacher replies, "When you know the rules, there's nothing hard about it."

The teachers note that most legends have a happy ending, but not this one. The children suggest various possible happy endings to the story. One boy's pen runs dry and he approaches me: "Hey, bro, can you give me a pen?" I do so. Another student tries to express himself, but finds it difficult to phrase a sentence.

The students are given a personal assignment relating to the text. They work hard. While they write, the teacher asks for help: "Can you guys tell me who's absent?"

One boy suggests: "Write down who's here and then compare the lists."

When the class renews its discussion on the moral of the story, one of the students quotes from the Jewish sources: "Money blinds the eyes of the wise and distorts the words of the righteous." The quote is a very accurate summary of the moral of the story. When I ask him where he knows this quote from, he replies: "From my counselor in the children's home."[9]

The teachers responsible for implementing the language and culture section of Kedma's curriculum face a difficult task. On the one hand, cybernetic discourse requires cooperation through dialogue; on the other, the teachers must project authority over a number of students who did not have an opportunity in their previous schools to acquire learning habits, enrich their language, and develop their emotional capacity to delay gratification. The conversation is calm and open as long as it is possible to control the students' behavior. When the balance is disturbed, the teachers do not hesitate to employ sanctions and forceful language in order to show the students that they are the boss. Despite this, the teachers are attentive to the students' comments and the students in turn listen to them and to one another. I did not witness any cases when a teacher spoke disrespectfully to a student who was unfamiliar with a given concept, however basic. In this respect, however, the children are less tactful than their teachers, as the following example shows:

Diary

A meeting of the student council. The council is discussing how to celebrate Purim (a Jewish festival with a carnival-type atmosphere). The teacher explains to the children that rather than hiring a prestigious discotheque, they could find an alternative venue and hire a disk jockey. One of the children asks, "What's a 'venue?' " The other children are irritated that he doesn't understand—"Somewhere to have the party, dummy!" The child is hurt and replies bitterly, "So why don't you just say somewhere for the party?!" When the language becomes too "colorful," the chairperson of the council looks at me, seems to recall why I am there and encourages her fellow students to tone things down.

"High" Culture and "Low" Culture

The tension between "high" and "low" culture can be found at all levels of the Kedma community. It is also evident in the teachers' narratives, as in the case of S's comments about the "right" music—Led Zeppelin, Pink Floyd, Shalom Hanoch, and others, compared to "socially bad" music such as the Mizrachi singers Nissim Saroussi, Zohar Argov, and so on. It is interesting that only those students who had abandoned any attempt at integration identified fully and shamelessly with the roughest manifestations of Mizrachi culture. S even commented that he used to take off the gold chains he wore[10] when he got to school.

These issues of "high" culture versus "low" culture, of embarrassment at the family's origin, and of discrimination were manifested when Professor Yehudah Shenhav[11] came to the school to meet the teachers. Shenhav gave his own personal examples of these issues.

Diary

Shenhav, a charismatic figure, smiles playfully as he recalls: "After we came to Israel from Iraq, my family also decided to change our name to something that sounds more Israeli. Once I appeared on television to talk about Israeli society. They asked what my name was. When I said Shaharabani, I felt as if I had said cunt and dick out loud to the whole nation."

In Kedma School, the prevailing style is one that consciously rejects "high" culture. The relations between teachers and students are highly informal, and ordinary slang is accepted. The subject of high and low linguistic registers is raised occasionally in staff meetings.

Ariella

Yesterday we had an argument . . . how to make the students "multilingual," in the sense that they can consciously and freely move from "high" language[12] to "low" language according to context, without feeling either inferior or patronizing. When they learn a richer and more intellectual language, they look for ways to keep their own identity and hold on to the language of the neighborhood as part of their identity. Some people gave Dr. Meir Buzaglo [of the Hebrew University] as an example of someone they admire. When he lectures on philosophy in the university, he uses the language of the neighborhood—he does not disguise himself or change his style. Others were disturbed by precisely this approach, and disliked the fact that when he lectures, he speaks like a "bro from the hood." Someone commented, "I don't like that pose." Rafi said: "I'm an existentialist, I don't want to compromise and lower my language. In the neighborhood I speak like people do in the neighborhood. I'm multicultural."[13]

Interestingly, however, the school also employs flourishes from "high" culture as a mark of cultural success. The annual production by the theater study track is an event that motivates the entire school, creating jealousy among less "prestigious" students. During rehearsals, the lead actors begin to show some of the mannerisms of stars, demanding (and receiving) "decent working conditions." Virtually the entire school community turns out for the final production, which is held on the Mt. Scopus campus of the Hebrew University of Jerusalem—a stronghold of Western cultural domination. The selected plays—American, European, and Israeli—address issues of oppression and alienation, such as relations between blacks and whites in the United States. Understanding the local context of these themes does not require much imagination. It should be noted that these themes are also addressed by the Israeli theater. The theater is a spiritual and almost religious bastion of the educated, secular section of Israeli society.[14]

The use of language, the concealment of Mizrachi traits, the use of the theater as a tool for nurturing cultural abilities, and the self-image of the children all testify to an interesting dialectic of attraction and rejection toward manifestations of "high" Western culture. In his meeting with the teachers at Kedma, Yehudah Shenhav discussed the blurring of lines separating identities, quoting his own example. After all, his power has been invested in him by "the bastions of Western Ashkenazi control." He noted that "identity is very complex. We are all Mizrachim [literally, Easterners], but we are probably all Westerners as well."[15]

Shenhav's comments echo views of McDermott and Varenne (McDermott & Varenne, 2002), who warn of the difficulty in containing within strict categories the increasingly complex identities that are emerging in a multicultural world. In Kedma, as in the other schools we visited, the search for identity is a complex and fascinating task.

Educational Discourse

As mentioned above, many teachers in the school draw on their personal narratives of discrimination in their work as educators, seeking to Repair their own experience for the sake of the coming generations. The term "Repair" is not intended to imply that they feel that there is something "wrong" with themselves; rather, the Repair involved has a social and political character—a Repair of empowerment, and an effort to find a voice and a place within a culture that has marginalized different populations.

The teachers do not adopt a one-dimensional approach promoting Mizrachi characteristics and denigrating anything that is Western or Ashkenazi. Rather, the school creates a critical discourse that confronts teachers and students alike with local and universal facts and narratives. The students develop their identity, in both the abstract and practical senses, by selecting from the range of cultural options with which they are presented.

Rafi, the philosophy teacher, and Hani, who teaches theater, explain why in their discourse with the students they use materials that are not usually identified with Mizrachi culture as anchors.

Diary

I observe Rafi's lesson. At first, some of the students find it difficult to relate to the subject of the lesson. Slowly, however, they are attracted by his skillful and interesting presentation of "Plato's Dinner." The students find similarities between the dinner and their own world. One girl seems worried. Rafi explains to me later that Plato's thoughts on love have an effect on her due to her own doubts about her relationship with her current boyfriend. Several students respond to Rafi's challenge and bring concrete examples from their own lives.

After the class, I question Rafi about the role and presence of Mizrachi philosophy in a school such as Kedma, which aims to foster Mizrachi conscience.

Rafi

I don't have to teach Mizrachi philosophy just because I am Mizrachi, although this year I have included some attention to the philosophy of Abrabanel. I think that Plato and Socrates are universal rather than Western or Eastern. I'm not going to stop liking Plato just because I'm Mizrachi.

Diary

Rafi discusses the difficulties faced by some of the students whom he supervises. The role of the "supervisor" in Kedma is holistic, combining aspects of the teacher, parent, psychologist and friend. He tells me about H, a student who was sometimes disruptive during the lesson I observed, but who also contributed some valuable comments to the discussion: "H tells me that when he gets bored, he goes out to steal motorcycle helmets. He comes from a very problematic background. I listen to him, and then try to work together with him to identify the disadvantages of the choices he makes. Recently I have begun to see some changes in him."

Chani

One of the things I do with the guys in the theater track is to go to plays with them. They like plays that are identified as having Mizrachi themes, but I have noticed that over time they have become more critical of the portrayal of the Mizrachi in the theater. They enjoy themselves, but they are also critical. They don't have any problem with characters identified as Ashkenazi, such as the characters in the plays of Hanoch Levin. They didn't like his plays because he is Mizrachi, but because he is a genius of the theater.

> They can relate to the pain that Levin expresses. I remember that we went to see a play by Yossi Banai called "The Life's Work." Sometimes the audience laughed but I didn't. A student sitting next to me asked me, "Why are they laughing? It isn't funny." The section was about missing out on things in life. I was very moved by the fact that they could come to the theater with the feeling that they are equal and take a critical stance.
>
> The students realize that the theater doesn't just belong to the "other guys" or rich people. Here they can meet themselves on different levels.
>
> Whenever we go to the theater, you can always hear people saying, "Who are those yobs?" I occasionally hear people making such comments, including students from other schools in Jerusalem. It really makes me mad. Once I accompanied another teacher who was taking her class to a play. When one of the students heard someone make this kind of comment, she said, "Who does she think she is, saying something like that?" I suddenly realized that things were much harder for us, because we didn't have Kedma School. We didn't even dare to think that the theater might be for us. The students at Kedma feel differently—they feel that the theater belongs to them, too.

Like many teachers at Kedma, Chani and Rafi have a profound and complex political, social, and philosophical worldview, based on a recognition that issues of identity are not one-dimensional or simplistic. The oppressed section of society cannot be empowered while denying or rejecting other sections. Whether they address history, philosophy, literature, or theater, the texts at Kedma are designed to expose the student and the teacher to universal and existential concerns. These texts—Mizrachi, Ashkenazi, European, and international—are reprocessed by the teachers and students, helping those who participate in the Kedma community to interpret their lives and their presence in Israeli culture and in the world at large. The teachers are often astonished by the comfortable and natural way in which the students become creative partners in addressing these texts, which were once considered the domain of restricted circles.

The empowerment of students through texts at Kedma is achieved primarily through the function of the teachers as cultural mediators, rather than neutral agents. The teachers actively encourage the students to internalize the existential and moral implications of the text for their everyday lives.

The development of a curriculum that emphasizes not only Mizrachi texts but also universal texts, from a standpoint of strength and empowerment, enables the students in Kedma School to gain from a cultural experience of strength, and to find their place within the multicultural discourse that takes place in Israeli society. The logic is that the school is producing a new generation of students capable, now and in the future, of engaging in struggles related to social justice and equality.

Integration in the Community

The school emphasizes the importance of its relationship with the surroundings and the need to influence the community. In the past, the students have had opportunities to prove that they have absorbed something of the school's critical and defiant approach. Kedma students have participated in demonstrations relating to various issues connected to the school and the community, have managed to advance issues and exert influence, and have gained considerable self-confidence through these activities.

> Clara
>
> I am jealous of them—I wish I could have gone to such a meaningful high school. One day the heating wasn't working. I spoke to the municipality but nothing happened. Later, the children organized a demonstration. I arrived from Tel Aviv and they told me about the incident. I supported them. The parents joined in the demonstration and the municipality responded. Within a day, the heating was working again. Olmert,[16] through one of his aides, called me and accused me of inciting the parents and youngsters, and claimed that I was behind the demonstration. That was the assumption made by the municipality. Our motto is to support the students on these kinds of issues.

The school is working very closely with the parents in order to rewrite the history of the Katamonim neighborhood through their personal narratives. There is no historical documentation of social action in the neighborhood. P, a teacher at the school, mentioned a fact when he presented the project to the parents: "I went to the Szold Institute to find material about the Katamonim, but everything was about crime, drug addicts and the Black Panthers."[17]

Encouraging the Parents' Narrative

In order to collect the "lost voices" that tell the history of the neighborhood, the school held a community evening during which the participants spoke of their recollections of the neighborhood and the relations between the residents and the authorities. The following are some of the comments that were made during the evening:

- They wouldn't let me use the postnatal services, because I didn't have fifteen Israeli pounds to pay.
- Moving into an apartment in the Katamonim after living in the transit camp felt like moving to a villa.
- When a car drove into the neighborhood, people used to ask who was sick—it must be the physician coming to pay a visit.
- I was a policeman. The pay was bad, but it was wonderful how everyone in the neighborhood helped each other. On Saturdays, everyone would sit on their balconies and tell stories. Everyone talked about where they had come from. We used to meet in the synagogue and then go on to drink arak. Sometimes people started to argue. I didn't like to get involved, because I live in the neighborhood, but when there was no choice I would tell one of them to bring some olives and the other one to fetch some arak, and they'd soon settle the matter.
- The school was not good, we spent all day learning the Bible, Torah and Talmud. In Arabic and English it took us a whole year to learn four letters. They didn't have secular studies.
- I didn't learn Gemara—I missed those studies.
- The schools weren't real schools.
- The schools were one big bluff.
- They didn't want to teach the "black" kids at school. They were second class, not part of the lexicon.
- I wasn't accepted to Denmark School, even though I passed the tests.
- My son was found to be gifted, but he was only accepted to a special program after I burst into [former Mayor] Teddy Kollek's office.
- I didn't know we were badly treated. Today I feel angry about it.
- The Panthers made a lot of noise, but they gave each of them a job to swallow up the movement and make them toe the line.

Kedma School does not confine itself to organizing projects such as the "Lost Voices" or documenting the history of the neighborhood. As part of its social education program, it also implements ongoing encounters with the nearby Arab neighborhood of Beit Safafa.

Clara explains that from time to time the youngsters met with teachers and students from Beit Safafa over a period of two years. She was amazed by the students' ability to learn to accept the other. The experience was far from simple, however. For example, the Jewish children heard comments such as "You guys expelled my family."

The school is currently seeking additional ways to influence its surroundings. The first class to graduate the school are members of an advisory council. As noted above, an effort is also being made to develop a "Kedma" model for a national educational program that would meet the needs of children from neighborhoods throughout the country.

Conclusion

S, a teacher in the school

The students at Kedma have fewer problems with their identity than in the past, although I agree that they still avoid showing pride in their identity in public. I am well aware of the stigma that exists—it is hard to ignore it. I do not think that matriculation grades will change this stigma, which is the result of geography, demographics and self-image.

When I tell people that I work in Kedma School, people treat me as though I have some kind of illness. My acquaintances assumed that I would stop working here once I stopped receiving special scholarships for this work. Even my colleagues shared that assumption. When I talk to people about the school, they ask me about the stigmas and about the children who come to the school. The stigmas have been so deeply internalized that even my father asked me whether the school has a future. But the children have an effect on me. Not all of them will manage to get a matriculation certificate, but I am sure that some of them will manage to complete their certificate four years later, as I did. We have strengthened their ability to stand up for their rights and their belief that they can make their own choices.

One of the graduates of the school spent her military service working with the families of injured soldiers. At a funeral she attended, she started to cry. Her commander was angry: "Their family mustn't see you crying." She stood up for herself, confident in her own strength and in her right and ability to change things.

Her commanders would not accept her approach, and in the end she left the position.

Is it our aim at Kedma that the children of our graduates will come to study here? That's the bottom line. As far as I am concerned, I would be very happy if my son chooses to study here. I hope I'll still be working here. We still haven't developed everything, but we have a chance to influence the processes that are shaping the school.

Of all the schools we visited, Kedma is the first that was not established on the basis of the ideology of encounter.[18] One reason for this is the failure of integration programs in education. While some students managed to succeed in integrative frameworks, although they were required to undergo a measure of cultural "conversion," many found themselves streamed into the lower strata of Israeli society. In many cases, the children of the elite retained their status, and children from poorer backgrounds remained at the same socioeconomic level.

Kedma School has been influenced by the pedagogics of liberation and by radical thinkers such as Freire (1994). The school aims to meet the needs of a large group of students who, after completing elementary school, found themselves labeled as failures, lost cases, and "problematic" students.

The school was established by professionals who developed a strong sense of identification with the students and the pain they carry. Many of the teachers relive the process by which they, as teenagers, were obliged to conceal part of their own identity as they attempted, not always successfully, to adapt themselves to their surroundings. This damaging process left them with a sense of being rejected. Not all the teachers share this experience, however; indeed, not all the teachers in Kedma School are Mizrachim or residents of the underprivileged neighborhoods.[19] These teachers come from a basic position of empathy. Some staff members do not have a highly developed social critique, but they confine themselves to making an effort to give their students a chance to succeed in the face the difficult challenges of global consumerism.

The goal of the school is that the students should be proud of their cultural heritage. It aims to counter the tendency of marginalized and oppressed groups to internalize oppression and to blame themselves for their situation. The school takes a holistic approach to the students, and the supervisor teachers provide a significant bond between the youngsters and the adult world. The supervisor is responsible for creating the conditions that will allow each student to succeed. The teachers in Kedma very rarely categorize their students or apply stereotypical and belittling approaches. Teachers who show signs of such an attitude usually find themselves leaving the school within a short period of time.

Despite the dramatic success rate of the first group of students in their matriculation examinations, despite the positive attitude of the teachers and their personal attention, and despite the devoted work of the staff, some families in the Katamonim neighborhood still see Kedma as a school for "disturbed" children. The students are not always proud to tell people that they attend the school. They and their parents ask, "If it is a good school, how come there are no entrance requirements?" These comments are evidence of the internalization of a negative self-perception (as in Woody Allen's version of Groucho Marx's famous remark "I'd never join a club that would allow a person like me for a member").

Kedma School is an example of what a devoted and idealistic staff can achieve with students who have been given up for lost by society. Once again, we see that attention, warmth, support, and, above all, love can enable almost any student to achieve. However, Kedma does not satisfy itself with just this. It also provides a political refuge in which educators with a highly developed social and political conscience can take part in a conceptual revolution through providing an education and acting as mediators of knowledge. This act of Repair is embodied in the establishment of an academic high school in neighborhoods that are usually home to vocational schools that are often uninspiring. The teachers engage in a constant and critical dialogue with themselves, and with the students and parents, with the goal of motivating the entire school community to take part in a process of empowerment and in the development of a positive identity. Thus the school makes an important contribution to social and critical discourse in still-emergent Israeli society.

The school provides space for harsh narratives of rejection and alienation that are largely the product of defective mechanisms of socialization in the State of Israel. Fifty years of living together should have led to significant social change, but the desperate need for a school such as Kedma shows that the lessons have not been learned. There is no shortage of theoretical sociological knowledge, but this has not led to even planning for reform and action, and the system continues to perpetuate the status quo.[20] As an aside, it is interesting to note that while the school in Neve Shalom more or less manages to exist as a model reflecting a genuine attempt to create an egalitarian encounter, there does not seem to be even a single school in Israel that seeks to create a similar integration between Ashkenazi and Mizrachi Jews. Mizrachim, at least for the moment, seem to have abandoned the hope of an option that would not demand that they eliminate and ignore their own culture.

CHAPTER FIVE

The Democratic School in Hadera

(1st–12th grades, approximately 300 students)

In their efforts to achieve a genuine and profound reform of teaching methods and traditional education, many educators in Israel and around the world have been influenced by the thought of Neill as reflected in his classic work Summerhill. This book, and the subsequent models developed by Kohlberg[1] and Greenberg,[2] have presented a serious challenge for those who advocate traditional and competitive educational methods.

During my stay in the United States, I worked in the complementary Jewish education system. This was a framework for evening studies attended by Jewish boys and girls who studied at various day schools. During the course of my educational work. I noticed that the boys and girls who came from a school based on a synthesis of the models developed by Neill and Kohlberg showed a higher level of leadership, creative thought, and assertiveness as compared to the other youngsters. These students seemed to have a simpler, freer, and more relaxed style. This also seems to be the impression created among those who visit the Democratic School in Hadera, one of the pioneers of this approach within the Israeli education system.

The Democratic School has a diffused and universal identity. Despite local characteristics, it forms part of an international chain that addresses broadly similar problems. In many countries, as in Israel, a tension can be seen between democracy and other forces active within society, with the result that many citizens are unaware of the true meaning of fully democratic life.

The founder of the Democratic School in Hadera also had a vision that related to a painful personal narrative from the past. Yaakov Hecht was dysgraphic and dyslexic, yet until the ninth grade managed to hide from his teachers the fact that he could not read. This astonishing story illustrates the way teachers are influenced by image, so that an intelligent student can fool a system that bases its evaluations on image. The school did not meet Yaacov Hecht's need for a special learning approach. After studying psychology, Hecht decided to establish the democratic school. In 1987, the school was established in cooperation with a team of educators, parents, and children. Three hundred students now attend the school, which is considered one of the largest of its type in the world.

Compared to the other schools we visited, every student in the Democratic School enjoys an extraordinary level of choice. Students can plan their own timetable and are not obliged to participate in lessons. However, once they begin to participate in a given course, they must meet its requirements. Several times a year, special workshops are run by the teachers, students, or outside experts.

The assumption is that as long as people are doing something that interests them without coercion or restriction, they will have a natural and proper desire to learn. In the regular system, this desire is often destroyed due to coercion, stagnant learning methods, and artificial order. At the Democratic School, there is no artificial pressure to "get through" a given quantity of material; it is accepted that each student has his/her own pace of learning.

> Teacher
>
> Some children are ready to read at age five, and some are only ready at seven or eight. So why does the system decide that everyone has to learn to read at age six? This decision means that half of all children are forced to become under-achievers. At least they didn't decide that all children in day care must learn to walk when they are one year old, because then they would have to give "remedial classes" in walking to fourteen month old babies.

> Diary
>
> The physical structure of the school is reminiscent of a kibbutz, like the school in Neve Shalom. The buildings are scattered over large areas, surrounded by eucalyptus trees. Little children play in the area around the junior section. Most of the young children are allowed to move around and play in areas that are a little further away from the junior section, but they mainly prefer to remain in "their" area.

Two six-year old children play with a friendly white cat. A child sits alone with a teacher, learning to read. Several children are working on the computer. Many of the teachers and children are dressed like Israeli vacationers in Sinai, in loose colored pants. Some of the children are unhappy about my presence. There are too many visits, and the school parliament, which meets every week and makes decisions on all aspects of life in the school, has been trying to put a limit on visits. When I approach, two boys launch into an "academic" discussion on radical education, as a humorous protest against my visit.

The ethos of the school is based on freedom. Each individual is entitled to do as they choose, as long as their acts do not injure other people or their environment. When students manage their own lives as free humans, they do not feel that education is being coerced on them. Even if a student does not do anything in school, they are still learning. If an eleventh grader angrily accused the school of "not forcing me to learn anything," the reply is, "he learnt a very important lesson—that it's not a good idea to waste time."

Democratic Management

The school does not have a principal in the usual sense of the word. I visited the school while Yaacov Hecht was still running it, and even then he was a member of the school parliament, with one vote just like everyone else! During the meeting I observed, he kept interrupting the chairperson, and after three warnings he was not permitted to speak for the remainder of the meeting. The discussion was about the need (or lack of need) to establish a laboratory in the school. Occasionally a six- or seven-year old student would try to make a comment, and the older students responded with tolerant smiles. The nonverbal message was apparent, though: the parliament is not a suitable forum for very young children.

The parliament is the legislature, while the executive arm (the committees) is responsible for implementing school policy according to changing needs. The judiciary exercises sanctions against members of the community who disobey the laws of the school.

Teacher

Sometimes students in the school push or hit one another, for example on the soccer field. But even the youngsters children know that if they warn an older child that they will go to the discipline committee, the older child is genuinely scared.

Despite the democratic approach, the school has a position for a principal, because the Ministry of Education does not recognize a school that does not have such a function. Nevertheless, there is a clear opposition in the school to all forms of centralized management. The various committees implement school policy (teachers' committee, admissions committee, finance committee, events committee, etc.). The flexible management and democratic process mean that the school is in a constant state of change.

Tenth Grade Student

All the students in the school should be involved in the parliament. We would like the situation to be perfect, but we aren't there yet. We are constantly changing things and asking questions. Our school asks questions. For example, the parliament is discussing the way students are admitted to the school, and after a day-long debate, the admissions process will be completely different than it is now. Every person who comes into the school changes something about it. And every person is also changed. If I tell you what's happening now, my comments will be conditional in nature, because I don't know everything, and what is happening right now will change in the future.

The idea is to integrate democracy in the school. There are two conditions for democracy: firstly, the presence of human rights, freedom of movement and thought, and so on. Here we also have the principle of equality between different age groups, something that you don't find in other frameworks. We are all equal. The second principle is the separation of powers. In a regular school, the teacher is the legislator, judge, critic and executor—the ultimate authority. Above the teacher you have the principal, and then people in the Ministry of Education that I don't know. Here, the powers have been separated. We have the parliament, the supreme authority to which everyone belongs. All the students, from age four to eighteen, as well as the parents and teachers, meet together once a week. Sometimes only twenty five people come for a discussion, but if there's an important issue, more come.

The children take their responsibility for the democratic management of the school very seriously. Accordingly, when they move on to other frameworks after the school, such as the army, higher education, or employment, they are not content to be cogs in the system. They grow up with the sense that they have the right to self-realization and are aware that those around them also have the right to freedom and self-realization.

An Institute for the Promotion of Democratic Education operates within the school, helping groups of parents to establish additional democratic schools.

Learning Style

Of the schools I visited, the extent of individual adaptation of learning styles in the Democratic School is the greatest. The free and noncoercive approach enables each student to find unique channels of expression and learning. In addition to formal activities, the students also learn to use informal time for learning, creating a very broad concept of learning.[3]

> Student
>
> In order to learn, you don't have to have a teacher, homework and exercise books. Learning can also take place by other means, such as through conversations and human interactions. That is why we have a lot of activities outside the classroom. Not far away from us, two children are sitting and drawing. It isn't an art lesson. When I wanted to learn how to draw portraits, I went to the art teacher, and he pointed me in the right direction and gave me homework. It wasn't a class assignment. The same thing applies to English or mathematics. There are different levels of math—it all depends where each student is.

The learning centers are open all day and are staffed by teachers who are familiar with the material. Each student can come and study without prior arrangement. The school also offers a variety of courses each week, which the students are free to join on a full or partial basis, or not at all.

> Diary
>
> I am near the library. A teacher is talking to two students she met.
>
> *Teacher*: What are you doing tomorrow morning?
> *Students*: Don't know, do you have plans?
> *Teacher*: If you come to the class, I'll come to teach.

Students: If you really insist, we'll come.
Teacher: If you really ask me nicely, I'll come.

Each student is required to choose one member of staff to serve as his/her mentor, helping to prepare the student's individual timetable, which has almost a university-like quality about it. If the open system we have described still fails to meet the student's needs, they can also reach an individual agreement with any teacher and study subjects not included in the curriculum. After negotiations, students can even spend several days a month "off campus" studying particular subjects.[4]

Contacts with the Community

Since this school has created a kind of artificial and democratically run polis (artificial since the members do not actually live there), the prevailing sense is that the school in itself constitutes a community. Unlike the other schools—Bialik, Kedma, and Neve Shalom—there is no geographical common denominator among the students, who come from as far away as Haifa, Tel Aviv, and Afula. The school could have been established in any community in Israel and, with various changes, anywhere in the world. It attaches less importance than the other schools to contacts with the immediate community, but it shows a strong desire to help other democratic schools wherever they may be. Many teachers travel around Israel to give lectures on the school, groups of interested parents visit the institution, and an initiative has been launched to establish an Institute for Democratic Education. The school seeks to influence others, and accordingly it publishes its ideas and helps establish additional frameworks that offer children a human and democratic educational option.

As of writing, the school does not have any organized framework of volunteering that requires personal commitment. Such activities took place in the past, and a teacher is currently examining this issue.

Many students wish to attend the Democratic School, but it is not easy to win a place. Teachers and students must face an admissions committee composed of teachers, parents, and students. The chances of a prospective teacher to be accepted will not be helped if they express surprise at the composition of the committee—teachers in regular schools are not used to seeing students in positions of authority.

In order to belong to the school community, prospective students must show a willingness to act within a democratic society, and the parents must support raising their children in the democratic spirit. The staff of the school believe that when there is a lack of congruence between the values manifested at home and those embodied by the school, the student suffers. Some students come to realize that they actually prefer a normative system based on competition and grades, in which they can strive to excel and achieve recognition. In such cases, they leave the school, sometimes against their parents' wishes. Yaacov mentioned a case when he had to persuade the parents to allow their child to leave the school, since he had a competitive nature and suffered from the lack of competitive challenges in the Democratic School.

> A Child Recalls,
>
> I wanted to join the school in the 9th grade, but there weren't any places available. When I started to slip in my school studies, I tried again, and that time I was successful.
>
> I stopped learning at my previous school because I disagreed with their approach. I became a poor student and wanted to attend an extraneous school. But in the end I found the best alternative. I will be taking the matriculation examinations.

Comments similar to these can be heard from many children who suffered in regular schools but are doing very well in the Democratic School. However, the admissions process and the culture of the school also reveal its weaknesses, which can be seen in other democratic schools in Israel. These schools tend to attract students of a certain type, most of whom come from secular, Ashkenazi, and left-wing homes. The students are not assessed according to these characteristics when they come to the admissions committee, or at least not overtly, but since there is a correlation between this profile of the student and the requirement of the school that the values transmitted at home should be "consonant with the spirit of the school," the school welcomes students who have already absorbed such values in the home.

Moreover, the parents at the school must be able to cover the costs that accrue from the fact that some of the children live at a great distance from the school and require transportation. These are expenses that can be met only by relatively prosperous families—monthly payments of around $125 are no small matter for many families.

Minister of Education Yossi Sarid may have been right in opposing the opening of a democratic school in Arad in 2000. There is some truth in the argument that democratic schools actually exacerbate social divisions, since they create concentrations of students from a particular and possibly elitist background and thus deprive other schools of these students.

The students at the Democratic School are aware of their cultural isolationism. At a meeting with students in the auditorium where the parliament had just met, three students from the school, including S, answered questions from trainee teachers. One trainee teacher asked about the way candidates are accepted by the admissions committee. "We only have selection to make sure that we don't accept anyone who's really out of line." S adds, "If they are rednecks or right-wingers, they're out of here right away."

From the standpoint of the ideology of Repair, this approach is problematic. While the school certainly inculcates democratic values among the students, from a sociological standpoint it appears to be preaching to the convert, catering to children who have been brought up in homes that inculcate specific cultural codes.

By contrast, the Experimental School in Jerusalem, which has adopted a radical educational approach (to varying degrees, over the years and during different periods)[5] has managed to maintain a measure of integration, as has Keshet School, which, as noted, was influenced by the Experimental School. A democratic (or radical, or free) school that operates in a disadvantaged area, or in a location enabling integration, should start from a different point of departure than that adopted by the Democratic School in Hadera. Kozol, for example, noted the weakness of the "free" model in the Afro-American ghettoes of the United States. These schools do not provide the children with basic skills such as reading, writing, and arithmetic—skills that Kozol feels are essential if these children are to survive in the technological world. He argues that the ghettoes need educated leaders with knowledge and degrees in architecture, engineering, and medicine, and, "whether you like it or not," such subjects cannot be learned spontaneously as one learns to walk or to make love.[6] Students in the Democratic School in Hadera have an advantage over students from poor neighborhoods. They have supportive families and they have benefited from informal cultural preparation for study and work. They are surrounded by "successful" people who can serve as role models. When I was involved briefly, in the early 1990s, in an attempt to establish a democratic school in a prosperous community, I was struck by the fact that all the parents were Ashkenazi, educated, and well-off, and it was this homogenous elitism, and the desire for social replication, that led me to leave the group.

I should note that, since I completed my research on the Democratic School in Hadera, Yaakov Hecht, the founder, while agreeing that there may be an elitist flavor to the Hadera school, informed me that there are two new successful democratic schools operating in poor neighborhoods in the greater Tel Aviv area. I heard this from other sources too, and it is certainly worthwhile analyzing and conducting research on these new efforts. My analysis here relates solely to the Democratic School in Hadera, and not to the new schools or any of the other sixteen or more democratic schools operating in Israel.

The Attitude to the Ministry of Education

As in the other schools I examined, this school also shows an ambivalent attitude toward the Ministry of Education. The position of principal would not exist in the school were it not for the demands of the Ministry of Education. The principal at the Democratic School in Hadera has less authority than principals elsewhere, and one of his main functions is to mediate between the school and the ministry.

The Democratic School has had its fair share of clashes with the educational authorities due to the differences between a democratic system and an authoritarian one. One interesting example is the initiative and work of a world-renowned artist who specializes in establishing playgrounds based on joint planning with children—an approach that is in keeping with the spirit of the school. When he agreed to come to the school to plan and build the play corner, there was much excitement among the students and teachers.

A student recalls,

> The children sold cakes, collected money and planned the play corner. They designed the area together with the visiting artist. But after we built it, the Ministry of Education ordered us to remove half the facilities because of various safety regulations. It's interesting that after we removed them, children began to get hurt.[7]

Nevertheless, dialogue does take place between the Ministry of Education and the school, mainly through the Education for Democracy Unit within the ministry.[8] An attempt has been made to make several existing schools more democratic. However, the Democratic School advocates the establishment of entirely new schools, due to the difficulty in introducing changes given the dominant culture in most schools.[9]

The Overt Curriculum

One of Kohlberg's sharpest criticisms of the regular education system related to the existence of a "hidden curriculum."[10] This term refers to all the aspects of the curriculum that do not appear in the declared policy of the school. For example, when I taught in a heterogeneous junior-high school in the Jerusalem area, everyone knew that the smoking corner was behind the gym, although smoking was officially prohibited throughout the school. There was a kind of covert agreement between the teachers on patrol during recess and the students that the latter would not smoke in public, and that the teachers would not follow them. Sometimes a student would come up to me straight after visiting the smokers' corner and say "How are things, teach?" We both realized that he smelt of smoke and it was obvious where he had been.

> Diary
> I visit the school in Hadera together with a group of interested parents. We see that a smokers' corner has been set aside for the joint use of students and teachers. Some of the parents are horrified to see a child who seems to be about thirteen with a cigarette in his mouth. The psychologist in the group comments: "It's better that he should smoke openly."

The democratic management of the school and the fact that the students take responsibility for their learning/life and are willing to discuss almost anything mean that almost any covert subject in the school immediately becomes overt and legitimate. Any subject is worthy of debate. When visiting trainee-teachers asked their host whether he was not afraid that the school may not prepare the children for "real life," he replied,

> Kindergarten is preparation for elementary school, and elementary school is preparation for junior-high. At junior-high they give the kids proper learning habits to prepare them for senior-high, and senior-high is preparation for the army and university. And university is preparation for life. At this school, we don't wait around. We are not preparing ourselves for life, we are living. Life is here and now.

This interesting comment reflects the desire to live life in the here and now, with all its complexity.[11] The fact that the school relates to life itself, rather than to a compartmentalized section of behavior, creates mechanisms for dealing with issues that remain covert in other schools. Moreover, since there are no tests or the pressures of competition, there is no need for the mechanisms of cheating that are common features of the hidden curricula of many schools.

In many schools, there is a substantial gap between slogans and reality, particularly in the fields of justice, equality, rights, and so on. The students learn that these are admirable slogans, but the school is sometimes run in a manner contrary to the slogans it professes, and by individuals who act in a forceful and discriminatory manner. Thus the students learn that there are "real and cruel rules of the game," and these constitute the hidden curriculum that lies behind the declarations, the school "ethos," and the "contracts" often signed by students in ordinary schools.[12] In the Democratic School, there are no empty slogans about democratic values. The students at the school do not live in an abstract manner. They learn about the practical meaning of democracy by grappling with real, everyday problems.[13]

Conclusion

The Democratic School in Hadera, established by a group of parents and educators under the leadership of Yaacov Hecht, a psychologist who was personally harmed by the trauma of "industrial" education, offers the regular education system an alternative form of education. This is an interesting option that could challenge the ordinary education system, which tends to scuttle a large number of students whose creative character is unsuited to the fixed form of the school. The Democratic School extends the concept of learning and applies it to life itself, attempting to create a "life here and now" for the community of those involved in the experiment.

Many students find at the school a response to their academic and personal needs that were not met in ordinary schools. Some are initially intoxicated by the freedom offered by the school and take time out for reflection and play. After a while, however, and without any coercion, they discover that they actually want to and like to learn. Sometimes, the students chose to study the very subjects in which they have difficulties, hoping to prove to themselves that they can succeed and in the process enjoying the challenge.

The children in this school learn to take responsibility for their own lives. This is reflected in the choice of the curriculum, in their participation in the democratic process that addresses almost every possible human issue, and in a pattern of study that combines individual preferences and the decisions of the majority, while taking into account the minority. This process of socialization fosters a more tolerant approach to the "other,"[14] though it is possible that children who come to this school come from homes that are already committed to such values.

In the school, excellence is defined not in terms of achievement relative to other students, but in terms of the individual's ability to set goals and to live and operate in accordance with his/her own principles and goals.[15] Some goals are set collectively, and hence achievement is at times collective. A large proportion of the studies takes place in mixed-age groups (a comparison of this field is offered below), both in informal learning through play and in the various programs offered in the school.

In cultural terms, and because of the profile of the students and teachers in the school, it has a highly universalistic and diffused identity. Many of those involved see themselves as citizens of the world capable of acting in accordance with the principles of universal justice. Many teachers are sent by the school to visit other democratic schools around the world, while others come to the school itself to study and spend time. These visits are not expensive, since the accommodation is provided mainly in the schools or in the teachers' homes. The school does not aim to socialize the students to any particular approach in terms of Zionism or Judaism. There is an atmosphere of academic freedom, and each person involved in the school community develops his/her own identity according to individual tastes and inclinations.

The school offers an interesting challenge for trainee teachers and for those already working in the field. The alternative education it offers is worthy of serious consideration, raising important questions. If the achievements of the students in alternative schools are no less impressive than those in regular schools, perhaps the choice should be reexamined. Do our children really deserve to suffer in a framework that can be strict, bureaucratic, impersonal, and arbitrary, when they could enjoy a framework that presents learning as a positive experience? In the United States, I met many students who had attended such schools, and they always tended to be more confident in making their own choices. In terms of personality, they were not afraid to take responsibility and play a leadership role within the group. When I taught in Jewish complementary education frameworks, I noticed that students who attended schools such as The Scarsdale Alternative School[16] almost always asked more complex questions and showed a greater level of interest than their peers.

At the same time, though, it is impossible to ignore the problematic aspects that accompany the development of a network of democratic schools. When parents exercise their right to give their children what they perceive to be the ideal form of education, the result is that students from certain social backgrounds are channeled to these schools, which thus perpetuate segregation and maintain distinctions between different sections of the population. The school is successful as an encounter of individuals, but it fails as an encounter of cultures, due to its homogeneity and the overt streaming of certain types of students to the school.

Yiftah Goldman, one of the founders of Tammuz urban kibbutz in Beit Shemesh, claims, "A democratic approach means that you have criticism of a given society and must struggle to change that society. Someone who creates alternative ghettoes abandons the basic principle of democracy, which calls for a change to society as a whole."[17] Goldman believes that Kedma School is far more "democratic" than the Democratic School in Hadera, since it enables young boys and girls to secure achievements that would otherwise have been closed to them. In so doing, it promotes greater equality within the education system in Israel, which Swirski argues is marred by institutionalized and deliberate inequality.[18] In his comments, Goldman reflects the interesting tension between democratic approaches that emphasize the principle of equality and those that emphasize the question of rights. The Democratic School is reminiscent of the American approach to democracy, which emphasizes personal rights and liberty. Goldman's criticism places him closer to the European approach, which finds fault with democratic societies in which economic and social gaps remain intact.[19]

It may be assumed that parents who choose to send their children to the Democratic School have complex reasons for doing so. In an article published on the Internet, Michael Lerner[20] calls for the examination of public education frameworks by comparison to alternative private frameworks. Lerner notes that it is easy to accuse parents who send their children to an alternative institution such as the democratic school of exploiting class privilege; however, these schools offer their students a chance to have a profound experience of the process of self-examination, to clarify existential questions, and to encounter more holistic thoughts on the world and human existence. These parents are disillusioned with the regular public system, which often fails to provide their children with a safe environment for emotional growth and the development of a sense of belonging. Lerner suggests that public systems would do well to internalize some of the methods employed by these special schools, which do not belittle the importance of providing a service for the broader community. The Democratic School in Israel could also serve as an important laboratory helping regular schools that seek to internalize and implement more democratic methods of learning. Accordingly, the Democratic School functions as a kind of mirror enabling the regular system to examine itself and critique its own actions.

With its ethos of choice and freedom, the Democratic School grants individuals the power to discover for themselves what approach they wish to adopt as they shape their own lives. In so doing, it addresses the questions raised by Neill in Summerhill,[21] as to whether parents have the right to mould the spiritual world of their children according to their own taste and beliefs and whether adults have the right to tell children how they should live. The answer provided by the Democratic School is clear: no. The freedom to decide within an educational system responsible for socialization, in a society that is still iconographic and highly particularistic in its approaches, creates an interesting tension that deserves further attention and study.[22]

CHAPTER SIX

Comparison and Conclusion

During the course of my research, I visited five schools that operate according to the ideology of Repair. Kedma School, the school in Neve Shalom, and Keshet School were all established with the express goal of repairing the rifts that exist in Israeli society. Bialik School underwent a transformation in order to address a new and emerging rift. Lastly, the Repair advocated by the Democratic School is more general in character, relating to the approach to study as implemented in "conventional" schools.

Myth and Ethos

As noted in our discussion on the model implemented in Neve Shalom School, a myth is a story that provides the members of a given community with their raison d'être, the meaning of existence of the group as such. The ethos is the way of life dictated by the myth. A myth consists of secondary myths that are connected to the metanarrative as told by the principal myth. Myths grow and die according to the needs of different groups at different times.[1] To this day, we employ Newton's theory as a functional theory, even though it has been proved incorrect, since, under certain conditions, it provides us with an adequate explanation. If we turn to the myth of Massada, we see that Josephus Flavius considered it a worthy story, since he was also a warrior, and hence exhibited bravery and heroism. For thousands of years, this myth lay "dormant" due to the disdain shown by Judaism and the rabbis for the mass suicide committed on Massada. Some eighty years ago, the myth sprang to life again due to the identification of the Zionist movement with the willingness of the warriors on the mountain to commit suicide rather than surrender or face capture. Massada became one of the secondary myths of the Zionist metanarrative, and Israeli soldiers climb the mountain and proclaim that "Massada shall never fall again."

Zionism is a unique national project of the twentieth century. It converted the mythological longing of the Jews for Zion into a practical ideology capable of conversion into reality. The first Zionists who came to Palestine saw themselves as partners in an ideology of Repair. They came to create a new society ex nihilo, to build a new kind of Jew who would not bend down before the Gentile but would be strong, assertive, and take control of their own destiny. Some, such as A.D. Gordon, attempted to maintain a symbolic connection with this Jewish tradition by modifying the significance of existing semantic codes. The word "work" (*avodah*) was transmuted from the ritual work (or worship) of God to working the land, as an act that develops individual spirituality no less than Divine worship does. Many of the founding generation were able to engage in a critical dialogue with the Jewish tradition, since, although they had rebelled against it, they were still thoroughly versed with its content. As time passed, however, and for various complex reasons, a gulf emerged between many secular Zionists and the Jewish traditions and texts. This was accompanied by a growing gulf between secular Israelis and most religious Israeli Jews, whatever their affiliation. Keshet School revisits this point of crisis, one of the symptoms of which is the creation of separate educational streams in the 1950s, and today it seeks to create Zionist discourse that brings students into contact with Jewish sources from diverse religious and secular standpoints.

Zionism provided a process of Repair for a people without a land who had paid a heavy price for living without sovereignty under the auspices of other nations. The national Jewish homeland was indeed established in the Land of Israel, but—contrary to the historical interpretation of many Zionists—this was not a "land without a people." The Repair of the injustice suffered by one people created a new injustice for another people. Tragically, the success of Zionism also constituted the point of departure of a new national rift between Palestinians and Israelis. This is the rift addressed in Neve Shalom.

The migration of millions of Jews from the Arab countries to Israel created a further rift of an ethnic and class nature, partly due to discrimination on the part of the establishment that absorbed the newcomers.[2] This establishment sought to impose a process of "cultural conversion" on the immigrants, on the basis of an ethnocentric belief that only one form of Jewish culture was deserving of replication. Opposition to cultural coercion was seen from the earliest periods of immigration, although it was only noticed by the authorities after the emergence of the Black

Panthers, the rise to power of the Likud, and the establishment of Shas. As we learned from the stories of the parents involved in Kedma School, their experiences as residents of the Katamonim were far more complex.

As an immigrant society with numerous rifts, the State of Israel is still seeking its identity and looking for a sense of belonging as part of the Middle East. The security threat makes it difficult for Israelis to develop an identity that does not relate directly to the political conflict and to a violent encounter with the "other." Despite these difficulties, the Israeli economy has grown and has come to accept almost without criticism the ideology of globalization and the free market, with all its disadvantages and costs. At the same time, Israel has become less centralized (due to the weakening of the Zionist metanarrative) and has witnessed the emergence of "communities of difference."[3] These communities are searching for their own voice, an identity within the sea of emerging and competing identities in Israeli society. While this trend implies a weakening of overall solidarity as a national concept, I also recognize that it brings a strong sense of solidarity and belonging within each of the communities of difference (Shas, immigrants, prosperous secular Israelis, etc.).

Each of the schools examined in this book lives according to a practical ideology of Repair manifested in concrete terms.[4]

The Formative Myth

Each of the schools I examined has a formative myth. In many cases, this myth relates to the distress faced by an individual who then motivated others, usually those who shared the same distress. An individual myth cannot motivate an entire community unless the potential to address the myth is already present. Based on his Christian ideology of Repair, Father Bruno Hussar[5] sought to repair the world by bringing Arabs and Jews to live together. He might have remained in splendid isolation, as have so many others in the region for thousands of years, had it not been for his success in motivating others who had been waiting—consciously or unconsciously—for just such an opportunity for Repair.

Clara Yonah, who suffered from the alienation of a culturally hostile environment, could have sat on the sidelines and waited with her vision. Instead, she managed to bring together a core group of leaders who joined her in an almost symbiotic fashion. Together, they developed the idea of a practical framework for a community of Repair.

Ruti Lehavi developed the ethos behind Keshet School, the central component of which is her own personal narrative as a religious woman living in a mixed religious-secular family. After the school was founded, it emerged that many parents from both "sides" were willing to engage in a process of dialogue that softens the myths held by each side. Religious and secular participants in Keshet School seek to free themselves of separatist myths that ignore the existence of the "other" as a partner in dialogue, action, and life.

Although Amira Yahalom did not establish Bialik School, she accepted it with all its complexity and found in it a place where she could develop an ethos based on love, acceptance, multiculturalism, and the goal of equality. The school itself became a myth, and has served as an example for all those who hope to develop a model for education in a multicultural society, particularly in poor socioeconomic backgrounds.

The principal and leading staff of each of these schools are familiar with their school's formative myth and recognize the changes that have occurred in the school and in its ethos due to various developments. The comments of both teachers and students provide echoes of these narratives.

In general, it is noticeable that these schools make extensive use of personal narratives as a key tool in understanding reality and constructing the school's complex ideology. The personal tragedy suffered by Boaz from Neve Shalom illustrates more powerfully than any academic theory the inherent difficulty faced by those who have chosen to reside in a community that lives the Arab-Israeli conflict on a daily basis, with all the advantages and disadvantages that this complex reality brings.

In each of these schools, the formative ethos is not totally binding, but it rather enables change in accordance with environmental and social developments. Although in most cases a single individual creates the initial myth and serves as its symbolic guardian, it is nevertheless understood and agreed that truth is created by the community as a whole and is not the sole prerogative of any individual.

Many of the stories told in these schools relate to an alternative interpretation of reality. The teachers use stories as an active tool for mediating between reality and the student. The quality of successful and convincing stories lies in their ability to offer a chance to improve reality. This explains the strength of the myths used in schools that operate according to the ideology of Repair. These schools realize the vision of Bruner, who speaks of the need to use narrative responsibly in order to teach students how to build the potential human—present, past, and future—in an interpretative manner.[6]

The School as Community

Since the Zionist myth began to weaken, Israeli society has begun to adopt unquestioningly the values of Western capitalism. This ideology impairs social solidarity and the sense of belonging of the nation's citizens. This is not a new phenomenon. When Durkheim[7] investigated the outcomes of the industrial revolution in France, he found an inverse correlation between the number of suicides and the strength of the connection between the individual and a socially cohesive community. Many people have been injured due to the rising level of alienation in Western consumer societies. This alienation leads many to seek alternatives to the social interaction that is gradually vanishing from the human landscape.[8] There has also been a decline in the level of public discourse relating to the search for common goals and values.[9] A growing tendency to nihilism accentuates the emergence of materialism and excessive consumerism as ostensibly desirable models for human behavior. Radical and extreme individualism, manifested widely in the media, politics, and economics, cannot meet the needs of the human spirit.[10] In my opinion, the atomization of Israeli society into communities of difference reflects a protest against this radical and materialistic individualism, among other causes.

The schools that embody an ideology of Repair are community schools in the fullest sense of this word: not as an empty slogan or a declaration in a long-forgotten document, and not as a sign on the front of the school, but as an existential ideology.

The growth of the Shas movement may be due in part to the failure of individuals to integrate into the Israeli myth of excellence, which nurtures many but leaves many others as failures facing a spiritual void and disturbed by feelings of failure and alienation. Shas gave people back their sense of belonging and provided spiritual meaning within a community framework. The schools described in this survey may be considered better than many others, including the Shas network, because they give their students a sense of belonging and meaning, while at the same time providing them with an education that enables them to find their place within a free, democratic, and multicultural society, in which technology, hi-tech, and science are vital features, and yet it does so while enabling them to maintain a critical stance to these characteristics.

The Community Characteristics of the Schools

The Bond between Children and Adults

The schools established on the basis of the Ideology of Repair create a genuine and close bond between students and teachers. It is rare that a student cannot find at least one teacher with whom he/she can enjoy a meaningful emotional and academic connection.

The Bond between the Children

Schools that implement the ideology of Repair strive to create a cohesive community that shares common goals and develops mechanisms for developing emotional and social skills. The teachers mediate between the children and encourage them to develop their own mechanisms for mediation.

The Bond with the Wider Community

Schools based on the ideology of Repair see themselves as connected with the community, with which they feel a profound affinity. They intend and plan to influence and change this wider community. Bialik School and Kedma School seek to change and improve their relations to the communities that surround their schools, in both the geographical and demographic senses. Keshet School is taking its first steps in this direction, absorbing a large number of children from the neighborhood and attempting to create integration on the basis of an ideology that sees heterogeneity as an advantaged rather than a burden. The school in Neve Shalom is the product of the local community, whose very existence embodies an attempt to influence Israeli society as a whole. The Democratic School sees itself as a pioneer in changing the Israeli education system, although it does not have an affinity to its geographical location.

The Bond with the Parents

Schools that implement the ideology of Repair see the parents as vital agents of socialization and strive to cooperate with them. The parents of students in Kedma School are invited to come to the school and relate the story of their neighborhood, thus helping to create an alternative source of information. The parents at Keshet School and the school in Neve Shalom are encouraged to join committees responsible for setting policy, while the parents at the Democratic School are full partners in managing the school. The parents at Bialik School find it difficult to come to the school since they work long hours, and their temporary status in Israel also creates a unique setting. Nevertheless, the school goes out of its way to enable meetings and cooperation, including publishing material in different languages. This approach reflects sensitivity to cultural differences. The school also arranges meetings late in the evening so that as many parents as possible can attend.

Acquaintance with the Children

Schools based on the ideology of Repair try not to let any child "fall between the cracks." Almost every student is recognized as a unique individual with his/her own needs, rather than as a number of a unit that creates grades. Staff meetings are often very long, since the problems of each child are discussed in depth. The teachers develop individual programs for each student based on the student's needs and learning styles. Every student is assigned a member of the staff who is usually made responsible for the overall development of the child in the physical, emotional, and intellectual realms.

Helping Children Bridge the Gap

Due to the populations they serve, Bialik School and Kedma School have paid particular attention to developing a policy of assistance that goes far beyond similar policies in regular schools. Resources are pooled with other sources of support in order to provide multidimensional assistance. This process can also be seen in Keshet School and in the school in Neve Shalom, albeit on a more limited scale, and on a still smaller scale in the Democratic School. The more prosperous the students' background, the less inclined the school is to take responsibility for areas that are usually the preserve of the parents or other systems.

The Sense of Belonging

Schools based on the ideology of Repair give their students a sense of belonging. One way they do this is by nurturing a sense of pride in the students' original culture and ethnic origin, in order to help them develop a positive self-image.

These schools are subversive in that they are community schools in a real sense of the term, and not as a hollow buzz word. They swim upstream in terms of the ethos of excellence (which favors some students but turns others into failures), and they offer the option of a cohesive community in an increasingly alienated and individualistic world. The existence and success of these schools constitute a voice of protest within Israeli society against the uncritical addiction to the values of materialism and excessive consumerism as a means of achieving happiness.

At the same time, differences can be seen between the various schools in terms of the perception of community, reflecting the different myths and formative ethoses of each school. All the schools may be positioned on the axis between the approach of Durkheim,[11] who sees the school as a place for the inculcation of values that already exist in society, and the approach of Dewey, who sees the school as a place for developing an understanding of the processes that can create a just society. The Democratic School would seem to be much closer to Dewey,[12] while Keshet and Kedma apply a critical approach that includes socialization to Jewish culture from unique standpoints. The school in Neve Shalom, which seeks to change the existing relationships between Arabs and Jews, works at the same time to reinforce the distinct identity of each group. Bialik School applies a more neutral approach, since it is still unclear as to what future it should socialize its students to that is, to which community they belong, since Israel is yet to develop a civil society. For the moment, Bialik confines itself to meeting the students' urgent existential needs. However, the very existence of the school, and the proposal currently under examination to establish additional schools of the same type, suggests the possibility in the future of a new community of difference that will be in no rush to leave Israel, and whose children will seek their own space and identity. Israel will have to encompass this community and develop a structured ideology regarding their life in the country. The present ideology is contradictory and cannot meet their needs.

If we accept the assumption that Jewish culture includes an inbuilt element of de-assertive attention to the "other" and to community in general,[13] the Democratic School can be positioned as the most individualistic of the schools examined in this study. Keshet and Kedma emerge as schools that strive to develop a communal Jewish identity based on values of the greater good, in idiosyncratic traditional contexts and with reference to traditional texts as a catalyst for examining the greater good and for criticizing society.

Developing an Innovative and Relevant Curriculum

All the schools examined in this study have shown themselves capable of developing new curricula that show a daring and interdisciplinary approach to knowledge. The findings from these schools are in keeping with the findings in Wood's study of innovative public schools in the United States. In his study, Wood found that these schools tend to "abandon" the textbooks and introduce studies based on "real books," with the adults playing a mediating role. The teachers develop a curriculum based on texts that shows internal order and coherence. The children use the cognitive tools they acquire in order to examine the world around them. Through their studies of language, history, and culture, they ask questions such as How can political power be secured? What is considered "cultural" and what "barbaric"? Who writes history? For whom and why? Wood found that many schools organize their studies around a central guiding theme, such as "Our roots in our school and our community."[14] Like the schools in Wood's study, the schools I visited use similar tools to inculcate in the students an ability to observe the world in a critical manner.

Cybernetic Knowledge

In all these schools, we find echoes of relativist theories relating to culture and to the production and representation of truth. Study is perceived not as a process by which empty vessels are filled with knowledge, but rather as an active process in which all those involved work as partners in the social structuring of knowledge.[15] The dynamic search for changing truth leads the schools to modify their myths and ethos. This is a highly complex process of dialogue that is not found in ethnocentric frameworks, which are threatened by the possibility of an encounter with "another" truth and a real encounter with the "other."

Learning Styles

The schools do their utmost to meet the unique styles of each student. Many ordinary schools also try to acknowledge different learning styles, but the large number of students in each class and the less community-like and supportive atmosphere impair the practical implementation of this ideology.

Multiple Voices

Schools based on the ideology of Repair enable multiple voices to be heard by providing space for tessellating, and sometimes contradictory, narratives. These multiple voices are consonant with the approach of Kohl, who recommends that students should feel comfortable with numerous narratives relating to themselves and others (Kohl's comments relate to women, Afro-Americans, Asians, and children of European origin). He argues that fairness in education can only be secured through study and discussion of a complex and sometimes painful history (his comments are made with regard to the United States, but are also highly relevant to Israel too).[16]

The Teachers in the Schools

The teachers in schools based on the ideology of Repair are more ideological than most of their peers elsewhere and tend to show a very high level of identification with their function. They are more open, more capable of operating in conditions of uncertainty, more willing to engage in an encounter with the "other" and to undergo a personal process of change while so doing. The creative and unique work that takes place in the schools of Repair seems to compensate these teachers for the low social status of their profession and their low salaries. These schools have the aura of a special community; teaching in them gives the surrounding community a sense that these educators are working out of a sense of mission and an ideological desire to achieve Repair.

The teachers are willing to work long hours, spend a long time preparing study materials, and attend exhausting meetings that focus on the progress of each individual student. The teachers receive positive feedback from their students and sense that their work influences their students' lives. Even teachers who began to work in these schools without any clear ideological direction internalize the ideology of Repair espoused by the school. Most of these schools nurture a new type of teacher who is culturally and socially aware and open-minded—qualities needed in order to face the changing needs of Israeli society. Despite the differences between the schools examined in this study, the development of the function of teachers in schools of Repair is quite similar.

Teachers who enjoy working in the ordinary education system dislike disorder and the absence of a fixed routine and disapprove of informal relations between teachers and students and so usually leave these schools and seek work elsewhere, particularly since such a profile is inappropriate for the ideology of the school. These schools are not suitable for teachers who have inflexible personalities.

The teachers in schools based on the ideology of Repair tend to throw themselves wholeheartedly into the learning process of the children. After all, as Wood reminds us, it is said of good teachers that they do not teach according to the curriculum, but they "teach who they are."[17]

Management

The schools of Repair are directed by teams that show an impressive capabilities in both practical implementation and in reflection. In some cases, the principals are also the founders of the school. The management teams are responsible for "telling the story of the school" in a broad sense of the word, both externally (to the media, politicians, and the Ministry of Education) and internally. They inculcate the existential mantras of the school among the teachers—not as hollow words, but as a concrete foundation for change and for improving human behavior in keeping with the school's ideology of Repair. The principals often use personal stories and narratives as a tool for improving the educational process. Teachers tend to adopt these stories and develop their own.

All the principals show a critical approach to the processes that are taking place in Israeli society. They do not accept the general political direction adopted by society. On the whole, they can be identified with a nonbourgeois, left-wing approach—a social left-wing that genuinely seeks to develop an appropriate program for social repair, equality, and the resolution of conflicts between opposing groups. The principals apply highly informal management methods, eschewing status symbols and self-importance. Their approach to others is direct and often humorous. The principals of schools based on the ideology of Repair represent a new generation of educators who are experts in education and bring an ideology of Repair to a divided and increasingly multicultural society.

It is still unclear whether these schools reflect an overall theory of multiculturalism in Israeli society. In other words, despite their common characteristics, it would not seem possible to amalgamate these five schools in order to establish a single framework encompassing the myths of all five. By way of example, it is difficult to imagine Keshet School working together with Bialik School or the school in Neve Shalom, due to their different positions on the question of identity.

Ramifications for Teacher Training

In his article "Multicultural, Intercultural Education—Can it Really Work?" Zvi Lamm notes that "people who are unable to develop a critical approach to the culture in which they live are unable to develop a tolerant attitude toward the culture of others."[18]

The schools we visited prove that when frameworks are established on the basis of the ideology of Repair, they do not wait to receive an appropriate ideology, but develop it themselves as a daily tool for interpreting and addressing problems. Teachers in Israel need to prepare themselves to meet the growing challenges of Israeli society—challenges that are already being addressed by these pioneering schools. In these frameworks, narratives serve as a key tool in developing the school's ethos. It is important that trainee teachers be exposed to the schools discussed here as part of their examination to test their adaptability to education in a changing and multicultural society. Teacher training colleges should also strengthen their contacts with these schools, possibly by integrating principals and senior educators from these schools as advisors on issues relating to social rift and multiculturalism. Exposure to these models constitutes a challenge and a catalyst for a critical discussion of educational issues in particular, and questions of power relations, inequality, culture, and socialization in general. In the future, teachers in Israel need to internalize and address these issues. As Stenhouse has already noted, "ultimately, it is the teachers who will change the world of schooling by themselves coming to understand it."[19]

This book seeks to engage those who work in the schools of Repair, teachers in other systems, and readers in general in a discussion on issues of education, identity, and socialization—not as philosophical abstracts, but as concrete experiences in light of the interesting practical actions we have observed, and in light of a practical ideology that seeks to make its voice heard and to offer a response to the problems faced by communities of difference. A familiarization with different voices, an understanding of what they have in common and where they differ, and, above all, a recognition of the importance of an encounter with the "other" per se may encourage the emergence of a social contract that has already begun to be implemented by these schools, giving us hope and inspiration.

What Next?

For some time, Israeli society has stood at a crossroads in its search for the best way to manage and overcome the challenges it faces. Some of the most complex challenges relate to the struggles that take place between different population groups over power, strength, representation, visibility, and values. Various groups organize around their own struggle, offering a remedy for the ills of Israeli society. Unfortunately, many Israelis have recently lost interest and hope in the possibility of securing

Comparative table of the schools

School/ aspect of comparison	*Neve Shalom*	*Bialik*	*Keshet*	*Kedma*	*Democratic*
Elitism	*Medium-high* high level of elitism among Jews. Among the Arab population there is successful integration across class lines.	*Very low* Children come from different backgrounds and share a low social status. Positive integration of different cultural backgrounds.	*Medium* The school absorbs a large number of students from the neighborhood. Who are successfully integrated.	*Very low* The school provides opportunities for children who have suffered due to their class status. The school integrates Mizrachi, Russian and Ethiopian students.	*Very high* There is no integration in the school, which effectively creates segregation due to its elitist separatism.
Encounter with the "other"	*Very high* The challenge is tough because of the nature of the national conflict addressed by the school.	*Very high* The challenge is tough because of the socioeconomic difficulties faced by the school population and the sense of transience.	*Very high* though within a single national group. An attempt is made to encounter other groups, including those outside the national group.	*High* The school emphasizes an encounter with the "other" from a position of cultural strength.	*High* In the context of the need to appreciate life with the "other" within a democratic society.

Continued

Comparative table of the schools

School/ aspect of comparison	*Neve Shalom*	*Bialik*	*Keshet*	*Kedma*	*Democratic*
Parental involvement	*Medium* Parents provide ideological support.	*Low* Characteristic of population of functional immigrants. Despite this, the school tries to develop links with the parents.	*Medium* Parents provide ideological support.	*Medium* School works hard to develop links with the parents.	*High* Ideological support.
Addresses specific rift	Arabs-Jews	Migrant laborers	Secular-religious	Failure of integration—class/ethnic rift	Not in concrete terms.
Ability to adapt to change	*High* Even minor decisions create difficulties and sometimes change positions.	*Very high* demographic changes occur on a daily basis.	*High* Adjustments are made to meet changing needs.	*High* Adjustments are made to meet the needs of adolescents who have had negative experiences in the education system in the past.	*High* Change is a function of the democratic nature of the school.

Integration in the community	*Very high* In the community; children who are bussed in from elsewhere are less well integrated.	*Very high* The school influences and changes the community.	*Medium* The school hopes to implement a higher level of integration.	*High* But the school aims for a still higher level.	The school itself is the community; no strong bonds with the surrounding community.
Attitude to the establishment	*Very critical* The mere existence of the school is an "act of repair."	*Very critical* The mere existence of the school is an "act of repair."	*Very critical* The mere existence of the school is an "act of repair."	*Very critical* The mere existence of the school is an "act of repair."	*Very critical* The mere existence of the school is an "act of repair."
Selected teachers	*Medium* Some teachers have a strong ideological commitment; others were just looking for work. The emphasis is on the process.	*Low* "Ordinary" teachers undergo an impressive process of change.*	*High* Word of mouth; contacts with teacher training college.	*High* Teachers come for ideological reasons; teachers with less strong connection undergo process.*	*Very high* Teachers share the school's philosophical approach.

* Process of change undergone in the school.

change through political activity—the forum for making the most important decisions in the state. Some do not believe that they will be able to influence the course of events; sections of the Israeli elite have developed a revulsion for politics and prefer to be active in other areas. As a result, particular attention is focused on the work of voluntary organizations that are attempting to encourage social, economic, and ideological change through grassroots work. These organizations, some of which take the form of social initiatives, engage in political struggles, but they generally refrain from using these struggles as a platform for entry into the world of party politics (examples of such bodies include the Democratic Mizrachi Rainbow, the Movement for the Quality of Government, and the Movement for a Constitution). The principals and educators committed to "Repair" whom we met at the schools we visited in this study belong to the category of leaders who seek a complete overhaul of social structures. In their own way, each of them attempts to address a given social challenge at the level of a new approach to education, seeking to create new individuals who, in the future, will be free of historical problems and failings and will acquire tools to resolve future problems by creative means.

The Repair of society by means of an encounter with the "other" is an important principle in the work of all the schools we visited, and it serves as a metaphor for the encounter with sections of one's self that usually remain dormant. These schools attempt to free their students from reactive thought, along the lines of "You did something to me, so I'll do something to you." Instead, they are offered a holistic cognitive approach that encompasses the existence, complexity, and rights of the "other" along with all their differences.

By promoting this approach, these schools attempt to dismantle the "monolithic Israeli identity," as Bar On (1999) calls it, that derives its force from the total negation of the "other," preferring a more "mature" process allowing the emergence of an identity capable of embracing diversity, multiple narrative, and complexity.

The education system should identify capable educators, of whom there are many, who work in the regular public education system and yet implement the ideology of Repair in their work. Locating such teachers and identifying forms of teaching that can foster excellence will help us find effective solutions for advancing disadvantaged multicultural and/or heterogeneous populations. Through this process, it will be possible to develop a bank of knowledge embodying an educational process of repair—something that is currently sorely lacking in the Israeli education system.

Similarly, we must identify within the regular public education system principals who act on the basis of the ideology of Repair, who can create foci of knowledge and practice relating to managerial excellence in Israel.

This study does not claim that only those schools that operate on the basis of the ideology of Repair seek to achieve meaningful change in Israeli society. Over the years, I have encountered a fair number of regular schools, with fine principals and teachers who advocate the ideology of Repair and do all they can to create meaningful change in the lives of their students. While these regular schools do not usually have a defined formative ethos such as those of the schools examined in this book, they nevertheless provide examples of educational actions that take account of cultural diversity and attempt to create social change by adapting the curriculum to changing circumstances and applying it. The following is the story of a regular Jewish public elementary school in a heterogeneous neighborhood of Jerusalem, close to the "seam line" between the Arab and Jewish areas of the city.

The principal of the school decided to plan an encounter between her students and Arab students from the neighboring village. She decided to use sport as the content of the encounter. Several additional events were later held relating to other themes. When the teachers met to summarize the experience, it emerged that many Jewish students felt uncomfortable because while the Arab students could explain the origins of their festivals, the Jewish students found it difficult, for example, to explain why they were collecting wood for the Lag Ba'Omer bonfires. They felt that they lacked information on Jewish heritage. With the encouragement of the principal, the teachers found ways to improve the quality of the encounters. More importantly still, joint discussion of the question of Jewish identity by teachers, parents, and students led the school to join the "Tali" (Enhanced Jewish Studies) track, which includes state schools that place a greater emphasis on Jewish studies and an acquaintance with the prayers, but from a pluralistic standpoint. This example clearly shows how an encounter with the "other" (Muslims, in this case) can lead to an awareness of overlooked or denied aspects of the "self" of the participants. Rather than abandoning the superficially "unsuccessful" encounters, the principal stood by her educational approach and initiated a process of self-examination that eventually led to a significant educational and cultural change in the school, by developing a new myth, and subsequently a new ethos, for the institution.

Like the principals we met in our survey, this principal was also able to read the reality in which she operated and seek alternative and relevant forms of interpretation. The structuring of the story of reality provides a tool for active mediation between reality and the child. Israeli society needs more leaders of this kind, leaders who have a vision or a convincing story, and the knowledge and skills to translate this vision or story into concrete reality.

Social Repair, the Ideology of Repair, and Social Entrepreneurship

The principals and educators we encountered are, I believe, social entrepreneurs who act in accordance with a clear vision of social Repair, or what we have termed an ideology of Repair.[20] In my opinion, this ideology contains remnants of the Jewish Socialist ethos that has undergone many changes over the years, (but has not completely disappeared) alongside a conscious rapprochement to some of the Jewish texts that discuss one's responsibility toward those who are different, weak, or alien.[21] At the same time, universal and Western influences can also be seen in this process, affecting the different schools in ways that are both common and distinct as they advance social entrepreneurship. At the international level, a growing body of literature documents analyzes and defines the phenomenon of social entrepreneurship. Social entrepreneurs may be found within each culture, in academic circles, and in diverse sections of the public at large. These entrepreneurs are defined as individuals who are capable of adapting and responding to innovation, effecting change, meeting new opportunities, and using resources from their environment (Davis, 2002). Drayton proposes several characteristics of social entrepreneurs: they achieve powerful and forceful systemic change, they are creative, the projects they establish have the potential to influence broad circles, and they have a strongly moral personality. In other words, the entrepreneur works in the world to create a society that is consonant with his/her vision. Accordingly, it is important that such entrepreneurs be able both to set goals and to solve problems. If either of these capacities is lacking, meaningful social change is unlikely. Entrepreneurs differ from ordinary professionals, who confine themselves to solving their clients' problems, in that they attempt to achieve broad-based and comprehensive change and are not deterred by difficulties and complex problems that take years to resolve. Neither do they hesitate to abandon aspects of their work that prove incapable of advancing their project. They show a healthy measure of realism (Drayton, 2002).

Bornstein notes that social entrepreneurs attempt to establish projects in specific areas, which are then extended after they secure initial success.

He comments that the outcomes of some of these initiatives are very impressive, even to the point of changing social policy. By way of example, he mentions Jeroo Billimoria, who founded Childline, a round-the-clock helpline for the 48 million children who live on the streets of India. The helpline was a tremendous success and led to the unification and enhanced efficiency of the services provided for these children. The success of this model led to its replication in three Latin American countries, and it continues to have a strong influence on street children in India (Bornstein, 2004). Bygrave notes that social entrepreneurs also seek to take responsibility for their own fate and to avoid dependence on their employers; their understanding of economic issues gives them a sense of ownership over their activities (Bygrave, 1994).

Around the world, there is an increasingly prevalent perception of social entrepreneurship grappling with the problems of migration, multiculturalism, inequality, and the class structure. The principals and educators at the "Schools of Repair" we visited are an example of such social entrepreneurship. Further investigation is now needed in order to discover additional schools that engage in social entrepreneurship and social change, and to learn more about instances of excellence and entrepreneurship within the "regular" system. Research and analysis should also focus on the activities of organizations that seek to advance an ideology of social Repair through problem solving and change—not only in the educational and academic sphere, but also in the fields of economy, culture, and politics. The potential educational value inherent in unravelling the actions of social entrepreneurs engaged in social repair is enormous. At the college where I teach, we launched a social entrepreneurship program in the 2003–2004 academic year, in cooperation with the Social Services Division of the Municipality of Jerusalem. The students in the program are expected to supervise and participate in various project and, after completing the four-year course, launch initiatives of their own.

After spending four months getting to know the different activities, the students were already more intensely involved in voluntary activities than they were in the past. After just six months, they even began to launch their own initiatives. There is something infectious about the enthusiasm of social entrepreneurs. We can reasonably anticipate that the more we investigate, analyze, and activate social initiatives, the more people will be drawn into the circle of those involved in such activities.

NOTES

Preface

1. See the studies of Bar-Navi and Ben-Ami (1998) and Ohana (1998) for a contemporary analysis of the rifts in Israeli society. See Almog (1997) for discussion of the disintegration of the myth of the Sabra.
2. The traditional Jewish term for this mending or repairing is Tikkun, translated here as "Repair."
3. See the study of Halabi (2000), which describes the working approach and experience he acquired in Jewish-Arab dialogue at the Peace School.
4. Established in 1974 as a progressive liberal party, Shinui made dramatic gains in 1999 after populist party leader Tommy Lapid remodeled that party around a strong anticlericalist platform.
5. Cf. Cohen and Zisser (2003), who noted that the results of a survey held in 5761 showed that the number of respondents who believed that the deepest rift in Israeli society was between religious and secular Jews was more than twice the number who believed it to be between Jews and Israeli Arabs.
6. I have formulated the concept "ideal of Repair" (Tikkun) over recent years as I have become acquainted with these schools. The concept was recently adopted by the principal of Bialik School as an efficient term for describing the work of the school.
7. One way to ascertain the reliability of findings in qualitative research is "triangulation"—the collection and cross-referencing of data from different sources prior to determining final conclusions (Elliot, 1991, pp. 80–83; LaBoskey, 1992, p. 178).

Chapter 1 Neve Shalom / Wahat Al-Salam

1. The diary sections of the book were written during or shortly after the visits. In other cases, I recorded impressions, thoughts, and descriptions of what I encountered.
2. Members of the minority group have developed various strategies for coping with their status. One approach is to attempt to integrate in the dominant group, whether as individuals or as a group. However, when it is difficult to move from one group to another, there is a heightened probability of a collective struggle to achieve an alternative and more just "social covenant." See: Suleiman (1999), pp. 171–186.
3. Benvenisti (1996), pp. 135–155.
4. Grossman (1992), p. 8.
5. Bar Shalom (1999), p. 146.
6. See Amir's table.
7. Mead (1974).
8. An elitist Arab school (junior high and senior high) attended by many Arab graduates of Neve Shalom after they complete elementary studies in the community.

9. The parents of children in the democratic school have already accepted the democratic approach. It would be difficult to establish such a school with parents who favor a more authoritarian approach.
10. Margaret Mead defines education as providing children with the tools for learning in the pre-figurative age (Mead, ibid.).
11. Amir (1969), ibid.
12. Literally: disaster (Arabic).
13. In a lecture at David Yellin College on the subject of multiculturalism, May 2001.
14. The Peace School runs binational workshops and seminars—see p. 31.
15. For more on this model, see Halabi (2000).
16. Who provided the inspiration for the film "Stand and Deliver."
17. Pedalino-Porter (1998).
18. See also the studies of Bekerman and Horenczyk, who describe at length the complex challenges that resulted from the opening of two bilingual schools in Jerusalem and the north of Israel (Bekerman, Horenczyk, 2001).
19. Neve Shalom deserves the compliments of Borstein (2001), who described the community as a "museum"—not of fossilized remnants of the past, but of future possibilities—and as a metaphor that has made itself come true.
20. Ohana (2000), p. 104.
21. Ibid., p. 107.

Chapter 2 Bialik School

1. Amira served as the principal of the school during the preparation of the study; she recently retired.
2. Maslow (1970).
3. Kozol (1992).
4. Harpaz (1985), pp. 31–32.
5. A description of the models may be found in: Greenberg (1992); Neill (1992).
6. Almog (1997).
7. Established in the early 1980s, Shas is an ultra-Orthodox party that has won widespread support among religious and traditionalist Mizrachi Jews.
8. It is interesting to note that at a convention on the subject of multiculturalism, Amira was criticized by the principal of a new school influenced by the Neve Shalom model. The principal, a Palestinian Israeli, attacked her for accepting the children of collaborators, whom he considers traitors to their people. Her response was: "If your humanism does not include these children, who is going to care for them?"
9. In the fall of 2004, the Knesset voted to grant full Israeli citizenship to the South Lebanese Army families living in Israel.
10. Amir (1969), chapter on Neve Shalom.
11. Bidol & Arima (1982). See also Herzog, who describes the disrespect and insensitivity that characterized the behavior of the establishment during the absorption of Ethiopian immigrants in the absorption centers. Bialik School presents almost the opposite example (Herzog, 1998).
12. Functional identification is found among immigrants when they do not have any ideological identification with their new country, but are prepared to give something (work, payment of taxes) in order to receive something in return (shelter, economic security); this as distinct from ideological migration, such as Jewish immigration to Israel for Zionist motives.
13. In some cases, teachers paid from their own pockets to cover the cost of ER treatment for children.
14. Gibton (2001), p. 546.

15. Amira and the other principals in this book seem to agree, on the whole, with Sergiovanni's claim that it is highly inefficient to mold school management after corporate world models (Sergiovanni, 2000). Unfortunately, in recent years this has become a popular (and, I would argue, very negative) trend in Israeli schooling, as also reflected in the Dovrat Commission.

Chapter 3 Keshet School

1. Since the text was written, Keshet School has opened its senior-high section.
2. For further discussion of secular-religious relations in Israel, see Liebman (1990).
3. Ruti Lehavi, the visionary behind Keshet School and its first principal, now directs the Association for the Promotion of Education in the Spirit of Keshet School. She is involved in curricular development, encourages the opening of frameworks similar to Keshet throughout Israel, and intends to return to the school to lead the high school which is currently being constructed as a continuation for the junior-high section.
4. A large number of organizations in Israel create encounters between Jews of different religious orientations. Melitz and Gesher are just two examples of bodies that organize various types of encounters.
5. See the discussion in Sheleg (2000) regarding the creation of a new form of Jewish religiosity that he terms "post-Orthodoxy."
6. Tali is a Hebrew acronym for "Enhanced Jewish Studies."
7. See a similar example in the study of Weil and Roer-Strier on Keshet School (2000), p. 43.
8. In a semiautobiographical study, Zimran (1991) describes the developments at the Experimental School during his term of office as principal.
9. Zimran, *What, You're a Child?*. op. cit., p. 280.
10. Amir (1969), in the chapter on Neve Shalom.
11. Erikson (1968).
12. Gershom Sholem, in Oroon (1998), p. 42.
13. Kasher (2000), p. 28.

Chapter 4 Kedma School

1. This situation closely resembles the experience in the United States, where streaming was used as an instrument for the desegregation of African Americans. See Wood (1993), p. 67.
2. Harrington (1979).
3. From a lecture by Avraham Hammou (2001), head of the Hammou Institute and formerly head of the Youth Division in the Israel Police. The findings quoted are taken from his MA thesis.
4. Delpin elaborated the concept of the culture of power. Those who have power deny its existence; those who lack power are more aware of its presence; Delpin (1995).
5. Bidol & Arima (1982). See also Landsman (2001), who emphasizes that minority students do not see reflections of their culture in the school setting (p. 65).
6. One characteristic of young people in this situation is a constant tendency to test limits, as if the student is trying to ascertain whether they will still be accepted even if they misbehave.
7. Ariella Bairy-Ben Ishay wrote a doctorate thesis on the work of the supervising teachers (Bairy-Ben Ishay, 1998).
8. Though the figures are unclear, the success rate for the first class of students in Kedma is 40 percent, compared to 6 percent in similar underprivileged neighborhoods. In many subjects, the success rate was close to 100 percent.

9. The reference is to a residential facility of the Youth Custody Service for children who cannot live at home. A number of students in Kedma live in such facilities.
10. Thick gold chains are considered a hallmark of the working-class Mizrachi male.
11. Professor of Sociology at Tel Aviv University who is himself of Mizrachi origin and has written on numerous subjects relevant to the issues addressed by Kedma School (trans.).
12. Delpin (1995) proposes that children should be taught to speak according to the appropriate cultural context in each given situation. For example, "public language" may be practiced through role-playing, drama, simulated newscasts, etc. In this way, the students learn that different languages are appropriate for different situations (Delpin, p. 53).
13. Nelson-Barber (1982) argues that there is a close affinity between language and cultural affiliation. Barber studied Native American students in the United States, and found that children in the 1st–3rd grades spoke a very similar dialect to that of their teachers. From the 4th grade, however, the children switched to the local dialect. Barber assumes that this is related to the children's desire to express their affiliation to an ethnic group to which they wish to belong.
14. From a lecture by Dan Orian at a conference on the subject of qualitative research, Tel Aviv University, September 10–11, 2001.
15. Meeting between Yehudah Shenhav and the teachers at Kedma, March 5, 2001.
16. Mayor of Jerusalem at the time.
17. The reference is to the Israeli "Black Panther," movement, named after the American Black movement. The Israeli Black Panther group emerged during the 1970s, demanding equality and justice for Mizrachi Jews, particularly those in underprivileged neighborhoods such as the Katamonim.
18. The school includes children from different ethnic backgrounds, but all share a low socioeconomic status.
19. Not all Mizrachim shared the experiences noted here, while some Ashkenazim also experienced rejection. While the ethnic divide is not completely indicative, it is statistically significant.
20. See Herzog (1998), who shows how the lessons of immigrant absorption in the 1950s were ignored during the absorption of Ethiopian immigrants at absorption centers in the late 1980s.

Chapter 5 The Democratic School in Hadera

1. Quoted in Power et al. (1989).
2. Greenberg (1998).
3. Indeed, the school states that the forum for learning is not the school as an institution, but life as a whole. The concept of learning is expanded to include "all the experiences an individual undergoes during their life," while the traditional approach included only "what happens in schools" (Harrington, 1979).
4. This approach reminded me of my Ph.D. studies in New York. We were permitted to build our own personal timetable, with the approval of our supervisor, and to take courses at Teachers College, Columbia University and other schools in the area.
5. For a detailed examination of this school, see Zimran, *What, You're a Child?* op. cit.
6. Kozol, in Cohen (1983), pp. 154–156.
7. An examination of his installations shows that they do not raise any safety concerns. The student who makes these comments speaks well and is self-confident. A brief inquiry reveals that he only started to read in 3rd grade.
8. The work of the unit has been dramatically curtailed recently due to policy changes.
9. For a detailed discussion of the difficulties encountered in creating change in schools, see Sarason (1996).
10. In Power et al. (1989).

11. The emphasis on the "here and now" is reminiscent of both humanistic psychology and the theory of Zen. The founder, Yaacov Hecht, was a psychologist. Although he himself has advanced degrees, he has noted that the high proportion of graduates of the Democratic School who have matriculation certificates and an academic education is a sign of "failure."
12. For an analysis of the phenomenon of the covert curriculum, see Power (1989), p. 21.
13. Power, Lawrence Kohlberg op. cit., p. 24.
14. See the study of Goldenberg (1998) on the impact of the school on the students' attitudes.
15. Greenberg (1992), p. 65.
16. Scarsdale Alternative School.
17. From a personal conversation (March 29, 2004).
18. See Swirski (1990, 1995).
19. An interesting finding may be noted in this context. When a comparative survey was carried out examining issues of freedom and equality as perceived by Americans and Europeans, the interviewers posed the following question: "Are you for or against government intervention in order to redress the gap in income created due to the free market?" An overwhelming majority of Americans (80 percent) were against such intervention, whereas some 60 percent of Germans supported it. Eighty percent of Italian and Austrian respondents agreed that the state should redress the imbalances created by the free market (Beck, 2002, p. 234). In Israeli society, the pendulum appears to have swung to the "American" side in an extreme manner.
20. An American publicist and thinker who is influential among left-wing social circles in the United States and exerted influence during the Clinton administration. Lerner is editor of the magazine Tikkun, which provides a platform for social criticism.
21. Neill (1992).
22. It should be mentioned that Yaakov Hecht has subsequently initiated the opening of 32 democratic schools in different parts of Israel, including disadvantaged neighborhoods.

Chapter 6 Comparison and Conclusion

1. See Kuhn's discussion of scientific revolutions (Kuhn, 1962).
2. Sternhal (1996) notes the possibility that the choice of an unequal "market" society was made at a relatively early stage in the development of the future State of Israel.
3. I owe the term "communities of difference" to Professor Yitzhak Kashti. Kashti has urged researchers to study and examine the communities of difference in order to enhance their awareness of the processes that led to the emergence of these communities. He argues that mechanisms of tolerance and solidarity are weakening, and the ability to reinforce these mechanisms depends on ensuring that each distinct community enjoys recognition and is able to make its voice heard (Kashti, 2001, pp. 3–4).
4. In this respect, they respond to the warning noted by Freire and Faundez of a disconnection between intellectuals engaged in abstract discourse and the actual needs of the populations they address (Freire & Faundez, 1989).
5. Bruno Hussar—the Christian priest who first suggested the idea of a joint Jewish-Arab community. Neve Shalom is the realization of his vision.
6. Bruner (2000), p. 101.
7. Durkheim (1951).
8. Dreyfus (1981).
9. Bellah (1981, 1985).
10. As well as Dreyfus (1981) and Bellah (1981), see also the convincing critiques of Lasch (1984), Peck (1987), and Sagan (1988) against consumerist and narcissistic culture as a chimera for the human condition. In their own ways, each of these scholars advocates a search for communal life and belonging in a world of disintegrating communal values. See also Etzioni (1994).

11. Durkheim (1956).
12. Dewey (1938).
13. Rottenberg (1983).
14. Wood (1993).
15. Kaney (2002).
16. Kohl (1994), p. 124; see also Bruner (2000).
17. Wood (1993).
18. Lamm (2000), p. 21.
19. In Rudduck (1998).
20. However, the specific interpretation varies from school to school.
21. In recent years, Amira Yahalom has combined her work in Bialik School with studies in Kollot, a Jewish studies program for secular educational leaders.

BIBLIOGRAPHY

Hebrew Publications

Almog, O. (1997), *The Sabra—A Portrait*, Am Oved, Tel Aviv.

Bar-On, D. (1999), *The "Others" in Our Midst: Changes in Israeli Identity from a Socio-Psychological Perspective*, Ben Gurion University of the Negev, Beersheva.

Bar-Shalom, Y. (1999), "*Mifgashim*—The Encounter Module between Diaspora Jewish Youth and Israeli Jewish Youth," in: A. Shai and Y. Bar-Shalom (eds.), *Qualitative Research in Education*, David Yellin College of Education, Jerusalem.

—— (2001), "Reactive Jewish Identity," in: *Bamikhlala* 13, David Yellin College of Education, Jerusalem.

Beck, U. (2002), *A Brave New World: A Model for a Civil Society*, Dark Red Line series, Hakibbutz Hameuchad, Tel Aviv.

Ben Jalloun, T. (1998), *Racism as I Explained to My Daughter*, Bavel, Tel Aviv.

Ben-Ammi, S. (1998), *A Place for All: Eli Bar-Navi Talks to Shlomo Ben-Ammi*, Hakibbutz Hameuchad, Tel Aviv.

Benvenisti, M. (1998), *A Place of Fire*, Dvir, Tel Aviv.

Borstein, D. (2001), "A Photo Lab," *Kol Ha'ir*, October 26, 2001.

Bruner, J. (2000), *The Culture of Education: Articles on Education in Context*, Sifriat Poalim, Tel Aviv.

Cohen, A. (1983), *Free Education*, Reshafim, Tel Aviv.

Cohen, A., and B. Zisser (2003), *From Acceptance to Escalation: The Religious-Secular Divide into the Twenty-First Century*, Shocken, Tel Aviv.

Criff, M. (1981), "Justice Is Theirs Alone," *Globes*, October 26, 1998.

Gibton, D. (2001), "The Leadership of the Principal in the Autonomous School," in: N. Tzabar Ben-Yehoshua (ed.), *Traditions and Streams in Qualitative Research*, Dvir, Tel Aviv.

Goldenberg, D. (1988), "The Influence of the Democratic School on the Knowledge and Attitudes of Its Students," MA thesis, Tel Aviv University, Tel Aviv.

Goleman, D. (1997), *Emotional Intelligence: The Special qualities of Successful People and How These Can Be Nurtured and Strengthened*, Matar, Tel Aviv.

Grossman, D. (1992), *Present Absentees*, Hakibbutz Hameuchad, Tel Aviv.

Halabi, R. (2000), *Dialogue between Identities: Arab-Jewish Encounters in Neve Shalom*, Hakibbutz Hameuchad, Tel Aviv.

Harpaz, Y. (1985), *The Community School: The Development of an Idea*, IACC, JDC-Israel, Jerusalem.

Herzog, T. (1998), *Bureaucracy and the Ethiopian Immigrants: Relations of Dependence in an Absorption Center*, Cherikover, Tel Aviv.

Horowitz, T. (1990), "Between a Neighborhood Community and an Ideological Community: The Founding and Institutionalization of Givat Gonen School in the Spirit of the Values of the Labor Movement," Henrietta Szold Institute (Internal Report), Jerusalem.

—— (1994), "Parental Choice of Schools as a Catalyst for Social and Value Change: An Ethnographic Study of Two Unique Schools with a Value-Based Character in Jerusalem," *Megamot* 36 (2–3), pp. 278–293.

Kasher, A. (2000), *Human Spirit*, Am Oved, Tel Aviv.

Kashti, Y. (2001), "The Interpretative Approach by Prof. Kashti," paper published following a lecture at the Mofet Institute, January 2001.

Keiny, S. (2002), "Community of Learners: Advancing Teachers and Making Them Learners," in: E. Shai and Y. Bar-Shalom (eds.), *Qualitative Research in Education: From Theory to the Field and from the Field to Theory*, David Yellin College of Education, Jerusalem.

Kohn, A. (2000), *The Education Our Children Deserve*, Hakibbutz Hameuchad, Tel Aviv.

Lamm, Z. (2000), "Multicultural, Intercultural Education—Can it Really Work?" in: M. Barlev (ed.), *Education for Culture in a Multicultural Society*, Department of In-Service Training, School of Education, Hebrew University, Jerusalem.

Neill, A.S. (2003), *Summerhill*, Keter and the Institute for Democratic Education, Jerusalem.

Ohana, D. (1988), *The Last Israelis*, Hakibbutz Hameuchad, Tel Aviv.

—— (2000), *A Humanist in the Sun: Camus and the Mediterranean Inspiration*, Carmel, Jerusalem.

Oron, Y. (1988), *Israeli Jewish Identity: A Study of the Attitude of Trainee Teachers from All Educational Streams to Contemporary Judaism and Zionism*, Papyrus, Tel Aviv.

Rogers, K. (1973), *The Freedom to Teach*, Hakibbutz Hameuchad, Tel Aviv.

Sheleg, Y. (2000), *The New Religious*, Keter, Jerusalem.

Shenhav, Y. (1998), "Human Dignity and Liberal Fundamentalism," *Ha'aretz*, December 10, 1998.

Shohat, E. (1995), "Toward Multicultural Feminism," *Noga* 28.

Souissa, A. (2001), "Explosive Transit Camp," *Kol Ha'ir*, October 26, 2001.

Sternhall, Z. (1995), *Nation Building or Social Reform?* Am Oved, Tel Aviv.

Suleiman, R. (1999), "The Collective Identity of the Palestinians in Israel," in: *Studies in Education* 4, Booklet 1.

Swirski, S. (1990), *Education in Israel: The Land of Separate Tracks*, Breirot, Tel Aviv.

—— (1995), *Seeds of Inequality*, Breirot, Tel Aviv.

Weil, S., and D. Roer-Sitrier (2000), *The Encounter between Religiosity and Secularity inside the School: The Case of Keshet School in Jerusalem*, Institute for Research on Nurturing in Education, Hebrew University, Jerusalem.

Yahalom, A. (5760), "The Children of Migrant Workers in Israel: A Proposal for an Educational Policy Framework at Bialik School" (paper prepared for the Knesset Education Committee), Tel Aviv.

Zimran, Y. (1991), *What, You're a Child?*, Carmel, Jerusalem.

Zucker, D. (1999), *We Secular Jews: What Is Secular Jewish Identity*, Yediot Acharonot, Tel Aviv.

English Publications

Adler, N. (2002), "Interpretations of the Meaning of Care: Creating Caring Relationships in Urban Middle School Classrooms," *Urban Education*, 37(2), 241–266.

Amir, Y. (1969), "Contact Hypothesis in Ethnic Relations," *Psychological Bulletin* (71).

Bairey-Ben Ishay, A. (1998), "Teacher Burnout and Consciousness-Complexity: An Analysis of the Mentors at Kedma (An Alternative Israeli High School," Doctoral thesis, The Graduate School of Education of Harvard University).

Banks, J.A., and C.M. Banks (1989), *Multicultural Education: Issues and Perspectives*, Needham Heights, MA: Simon & Schuster.

Bellah, R. (1981), "Cultural Vision and the Human Future," in: D. Sloan (ed.), *Toward the Recovery of Wholeness*, New York: Teachers College.

Bellah, R. (1985), *Habits of the Heart*, Berkeley, CA: University of California Press.

Bekerman, Z., and G. Horenczyk (2001), "Bilingual Education in Israel," Report Submitted to the Ford Foundation and the Center for Bilingual Education in Israel.

Bennett, C. (1990), *Comprehensive Multicultural Education*, Boston: Allyn & Beacon.

Bidol, P., and R. Arima (1982), "Multicultural Perspectives in Human Relations Training," in: L. Porter and B. Mohr (eds.), *Reading Book for Human Relations Training*, Alexandria, VA: NTL Press.

Bornstein, D. (2004), *How to Change the World: Social Entrepreneurs and the Power of New Ideas*, New York: Oxford University Press.

Bowels, S., and H. Gintis (1976), *Schooling in Capitalist America*, New York: Basic Books.

Brislin, R., and T. Yoshida (1993), *Intercultural Training: An Introduction*, Thousand Oaks, CA: Sage.

Bygrave, W.D. (1994), *The Portable MBA in Entrepreneurship*, New York: John Wiley & Sons, p. 5.

Davis, S. (2002), "Social Entrepreneurship: Towards an Entrepreneurial Culture for Social and Economic Development." Paper presented to The International Board Selection Committee, Ashoka: Innovators for the Public.

Delphin, L. (1995), *Other People's Children: Cultural Conflict in the Classroom*, New York: New Press.

Dewey, J. (1938), *Experience and Education*, New York: Collier Books.

Drayton, W. (2002), "The Citizen Sector: Becoming as Competitive and Entrepreneurial as Business," *California Management Journal*, 44(3).

Dreyfus, H.L. (1891), "Knowledge and Human Values: A Genealogy of Nihilism," in: D. Sloan (ed.), *Towards the Recovery of Wholeness: Knowledge, Education and Human Values*, New York: Teachers College.

Dunn, R., and S. Griggs (1988), *Learning Styles: Quiet Revolution in American Secondary Schools*, Virginia: NASSP.

Durkheim, E. (1951), *Suicide*, George Simpson (ed.), Glencoe, IL: Free Press.

—— (1956), *Education and Sociology*, Glencoe, IL: Free Press.

Elliot, J. (1991), *Action Research for Educational Change*, Philadelphia: Open University Press.

Erikson, E. (1968), *Identity, Youth and Crisis*, London: Norton.

Etzioni, A. (1994), *The Spirit of Community: The Reinvention of American Society*, New York: Touchstone.

Feurverger, G. (1995), "Oasis of Peace: A Community of Moral Education in Israel," *Journal of Moral Education*, 24(2), 113–141.

—— (1997), "An Educational Program for Peace: Jewish-Arab Conflict," *Theory into Practice*, 36(1), 17–25.

Freire, P., and A. Faundez (1989), *Learning to Question*, New York: Continuum.

—— (1994), *Pedagogy of the Oppressed*, New York: Continuum.

Gardner, H. (1992), *Multiple Intelligences: The Theory in Practice*, New York: Basic Books.

Gillman, N. (1990), *Sacred Fragments: Recovering Theology for the Modern Jew*, New York: JPS.

Greene, M. (2002), *Variations on a Blue Guitar: The Lincoln Center Institute Lectures on Aesthetic Education*, New York: Teachers College Press, p. 65.

Greenberg, D. (1992), *The Sudbury Valley School Experience*, Framingham, MA: The Sudbury Valley School Press.

—— (1998), *Reflections on the Sudbury Valley School Concept*, Framingham, MA: The Sudbury Valley School Press.

Harrington, C. (1979), *Psychological Anthropology and Education: A Delineation of a Field of Inquiry*, New York: AMS Press.

Hersh, R.H., D.R. Paolitto, and J. Reimer (1989), *Promoting Moral Growth: From Piaget to Kohlberg*, New York: Longman.

Holt, J. (1971), *Freedom and Beyond*, New York: Penguin Books.

Kohl, H. (1994), *"I Won't Learn from You" and Other Thoughts on Creative Maladjustment*, New York: The New Press.

Kozol, J. (1992), *Savage Inequalities: Children in America's Schools*, New York: Harper.

LaBoskey, V.K. (1992), "Case Investigations," in: Judith H. Shulman (ed.), *Case Methods in Teacher Education*, New York: Teachers College Press.

Landsman, J. (2001), *A White Teacher Talks about Race*, Lanham, MD: Scarecrow Press.

Larrivee, B. (2000), "Creating Caring Learning Communities," *Contemporary Education*, 71(2), 18–21.

Lasch, C. (1984), *The Culture of Narcissism*, New York: W.W. Norton and Co.

Lerner, M. (2002), "Don't Let the Lights Go Out," A post-election op-ed about dealing with post-election depression (distributed via e-mail by magazine@tikkun.org).

Levy, S., and E. Katz (1993), *Beliefs, Observances and Social Interactions among Israeli Jews*, Jerusalem: Israel Institute of Applied Social Research.

Liebman, S.C. (1990), *Conflict and Accommodation between Jews in Israel*, Jerusalem: Keter.

McDermott, R.M., and H. Varenne (2002), "Reconstructing Culture in Educational Research," (Unpublished paper).

Maslow, A.H. (1964), *Religions, Values and Peak Experiences*, New York: Viking Press.

—— (1968), "Some Educational Implications of the Humanistic Psychologies," *Harvard Educational Review* 38.

—— (1970), *Motivation and Personality*, New York: Harper.

Mead, M. (1974), "Grandparents as Educators," in: H.J. Leichter (ed.), *The Family as Educator*, New York: Teachers College Press.

Merriam, S.B. (1991), *Case Study Research in Education: A Qualitative Approach*, San Francisco, CA: Jossey-Bass.

Neill, A.S. (1992), *The New Summerhill*, London: Penguin.

Nelson-Barber, S. (1982), "Phonologic Variations of Pima English," in: R. St. Clair and W. Leap (eds.), *Language Renewal among American Indian Tribes: Issues, Problems and Prospects*, Rosslyn, VA: National Clearinghouse for Bilingual Education.

Noddings, N. (1992), *The Challenge to Care in Schools: An Alternative Approach to Education*, New York: Teachers College Press.

Olsen, L., and M. Chen (1987), *Crossing the Schoolhouse Border: Immigrant Students and California Public Schools*, San Francisco, CA: California Tomorrow.

Pang, V., J. Rivera, and J. Mora (1999), "The Ethic of Caring: Clarifying the Foundation of Multicultural Education," *The Educational Forum*, 64(1), 25–32.

Pearce, P.L. (1982), "Tourists and Their Hosts: Some Social and Psychological Effects of Intercultural Contact," in: S. Bochner (ed.), *Cultures in Contact: Studies in Cross-Cultural Interaction*, Oxford: Pergamon.

Peck, M.S. (1897), *The Different Drum: Community Making and Peace*, New York: Simon and Schuster.

Pedalino-Porter, R. (1998), "Why Even Latino Parents Are Rejecting a Program Designed for Their Children's Benefit," *The Atlantic Monthly*, May 1998.

Pedersen, P. (1994), *A Handbook for Developing Multicultural Awareness*, 2nd edition, Alexandria, VA: ACA.

Power, C., A. Higgins, and L. Kohlberg (1989), *Lawrence Kohlberg's Approach to Moral Education*, New York: Columbia University Press.

Rogers, C. (1961), *On Becoming a Person*, Boston: Houghton.

Rudduck, J. (1998), "Changing the World of the Classroom by Understanding It: A Review of Some Aspects of the Work of Lawrence Stenhouse," *Journal of Curriculum and Supervision*, 4(1), 30–42.

Sagan, E. (1988), *Freud, Women and Morality*, New York: Basic Books.

Sarason, S.B. (1996), *Revisiting "The Culture of the School and the Problem of Change,"* New York: Teachers College Press.

Sergiovanni, T.J. (2000), *Leadership for the Schoolhouse: How Is It Different? Why Is It Important?*, San Francisco, CA: Jossey-Bass.

Sleeter, C.E., and C.A. Grant (1988), *Making Choices for Multicultural Education: Five Approaches to Race, Class and Gender*, Columbus, OH: Merrill Publishing Company.

Sparks, D. (2004), "From Hunger Aid to School Reform—An Interview with Jerry Sternin," *Journal of Staff Development* 25(1), 46–51.

Tatum, B. (1997), *Why Do All the Black Kids Sit Together?*, New York: Basic Books.

Taylor, C. (1992), *Multiculturalism and "The Politics of Recognition": An Essay*, with commentary by Amy Guttman (ed.), S.C. Rockefeller, M. Walzer, and S. Wolf, Princeton, NJ: Princeton University Press.

Valdivieso, R. (1990), *Demographic Trends of the Mexican-American Population: Implications for Schools*, Charleston, WV: ERIC Clearinghouse on Rural Education and Small Schools.

Willshire Carrera, J. (1989), *Educating Undocumented Children: A Review of Practices and Policies* (Trends and Issues Paper), Charleston, WV: ERIC Clearinghouse on Rural Education and Small Schools.

Wood, G.H. (1993), *Schools That Work: America's Most Innovative Public Education Programs*, New York: Penguin-Plume.

INDEX

Educating Israel is a comparative ethnography of five Israeli Schools that operate within a multicultural context, offering an important glimpse of the individuals who bravely act as social and educational entrepreneurs striving to change Israeli society from within. In all five of the schools Bar Shalom examines, there is an educational emphasis on accepting the Other, regardless of his or her ethnic background and religious affiliations. This book demonstrates how Arab, Jew, Moslem, Christian, secular, religious, and Mizrachi students can come together to become agents of change in Israel's conflicted and pluralistic society. Bar Shalom shows us how these state-of-the-art schools, with their innovative curriculums and philosophies, are more than institutions of learning, serving as models for change in Israeli education and in Israeli society as a whole.

Yehuda Bar Shalom has many years of experience as teacher and educator, both in formal and informal settings. He is a member of the Arab-Jewish dialogue group at David Yellin College, a co-founder of the Social-Educational Entrepreneurship network, in conjunction with Jerusalem's municipality, and a co-editor of the textbook *Qualitative Research in Education: From Theory to Field—From Field to Theory*.

GPSR Compliance
The European Union's (EU) General Product Safety Regulation (GPSR) is a set of rules that requires consumer products to be safe and our obligations to ensure this.

If you have any concerns about our products, you can contact us on

ProductSafety@springernature.com

In case Publisher is established outside the EU, the EU authorized representative is:

Springer Nature Customer Service Center GmbH
Europaplatz 3
69115 Heidelberg, Germany

www.ingramcontent.com/pod-product-compliance
Lightning Source LLC
LaVergne TN
LVHW100315210926
848480LV00003B/1123
9781349533954